I0064235

MARKETING OF SPICES

MARKETING OF SPICES

Vigneshwara Varmudy

2016
Daya Publishing House®
A Division of
Astral International Pvt. Ltd.
New Delhi - 110 002

© AUTHOR
First Impression, 2001
Reprinted, 2016

ISBN: 978-93-5130-983-3 (International Edition)

Published by   :   **Daya Publishing House®**
                      A Division of
                      Astral International Pvt. Ltd.
                      – ISO 9001:2008 Certified Company –
                      4760-61/23, Ansari Road, Darya Ganj
                      New Delhi-110 002
                      Ph. 011-43549197, 23278134
                      E-mail: info@astralint.com
                      Website: www.astralint.com

Laser Typesetting   :   **Goswami Printers**
                      Delhi - 110 035

Digitally Printed at   :   **Replika Press Pvt. Ltd.**

PREFACE

India has a worldwise reputation as the only country which produces almost all kinds of spices. Spices are virtually indispensable in the culinary art and most of them are actually the housewife's secret of tasty dishes. These spices also play a significant role in the development of Indian economy. It is through these varieties of spice's export that India has earned the much needed foreign exchange over a long period of time. The spices actually determine the health, wealth and policies of the nation.

Several spices have originated in India viz., Black Pepper, Cardamom etc. India ranks first in terms of area and production of black pepper, large cardamom, ginger and turmeric. India once had the monopoly in the trade of spices in the global market but is losing out it's grip in recent years. Despite the tremendous importance of spices, it is rather unfortunate that our policy makes have not succeeded in improving this sector. There appears to be a prevalence of several hurdles in this sector, which actually demand certain planned solutions so as to develop this in particular and the country in general.

It is in this regard, that the present volume tries to highlight some of the important economic as well as other aspects of the spices of India. This volume tries to deal with the various aspects of some of the major spices like Black pepper, Cardamom, Ginger, Turmeric etc., along with some other minor spices. A special focus has been given to the problems of production and marketing of spices along with some of the measures to promote this industry. Eventhough a lot of work has been undertaken on spices by the researchers, administrators and institutions, most of these works are scattered. Again, the author of this volume felt that it is very rare to see almost all the aspects viz., production, uses, marketing, problems etc., in a single volume with required data. Hence, this little effort is being made.

The present volume gives the required available statistical data so as to support the views. While giving this, the author had to heavily fall back upon a number of previous works, research articles, reports, journals, bulletins etc., for which the author feels grateful to all concerned.

While giving information on various spices, in the initial chapters importance is being given on those spices which are playing an important role in the development of the economy at present, and an attempt is also made to highlight the importance of the so called minor spices which are more or less neglected, but they are having a bright future and can build the sector in the long run. At last, by highlighting the problems of the spice industry along with certain measures, the author also feels happy to give the information on the role of organisations in the development of this vital sector.

I express my deep gratitude to the management, staff and friends of Vivekananda College, Puttur for their advice and help in bringing out this publication.

I am highly indebted to my principal Prof. A.V. Narayan, for his advice and suggestions, which were very fruitful to me in the completion of this volume.

The association of my wife Mrs. Poornashri in my academic pursuits is unforgetable to me. My duty for acknowledgement will not complete if I don't acknowledge the cooperation of my children Shreyaswi, Nithin Shankar and Sachin Keshava.

29-03-2000   DR. VIGNESHWARA VARMUDY

CONTENTS

LIST OF TABLES

(x)

1

INTRODUCTION

India is the land of spices. As the country is blessed with a wide range of agroclimatic conditions from tropical to temperate zones, coastal plains to high altitudes and semi-arid to highly humid evergreen forests, therefore, it is in an advantageous position to produce a number of spices. As per the terminology on spices prepared by the Bureau of Indian Standards, there are sixty three spices produced in this country. The spices grown in India may comprise different plant components or parts such as Floral parts (Cloves, Saffron etc.) or Fruits (Cardamom, Chillies etc.) or Berries (Black Pepper) or Seeds (Celery, Coriander etc.) or Rhizomes (Ginger, Turmeric) or Roots (Tejpet) or Leaves (Mints) or Arial (Mace) or Bark (Cinnamon) or Bulbs (Garlic) or other parts of spice plants.

The history of spices in India was started, perhaps, with the beginning of human civilization of this country. No country in the world grows as many kinds of spices as India because of this India is known the world over as "The Home of Spices". The fame of Indian spices is older than recorded history. The fascinating history of spices, is a story of adventure, exploration, conquest and fierce navel rivalry. In a nutshell, battels have been fought over them, civilization grew around them, colonization, birth of monopolistic discovery (The East India Company) and the discovery of the 'New World' by Columbus are all products of the quest of spices. Today, India has a unique position in the global spice scenario as the largest producer, consumer and exporter.

Among the different varieties of spices grown in India, the important among them are black pepper, cardamom, ginger and turmeric that were originated in this country. Spices such as chillies, clove, nutmeg and vanilla were introduced by the invaders. The black pepper is called as 'The King of Spices', Cardamom as the 'Queen of Spices' and turmeric is once known in the West as 'Indian Saffron'. Both pepper and cardamom are believed to be originated in the Western Ghats of the country. The other major spices such as ginger, turmeric and chillies are grown here to a large extent. South India commands sole or near monopoly in the production of pepper, ginger and turmeric; there are a number of other spices which are contributed from North India in sizable quantities.

In terms of area and production of pepper, cardamom and ginger Kerala stands first, while Andhra Pradesh has it's dominance in turmeric and chillies. Rajasthan is the main producer of coriander, cumin and fenugreek. Madhya Pradesh is popular for garlic. Sikkim possesses maximum area under large cardamom and in production too. (Table-I) The area under clove is maximum in Tamil Nadu and Kashmir is the major producer of Saffron in India.

TABLE-I : Area, Production, Productivity and States which Stand First in the Respective Spices in 1997-98

Spices	Area ('000 ha)	Production ('000 tonnes)	Productivity (kg./ha.)	Name of the State in area and production
Black Pepper	181.55	57.27	315	Kerala
Cardamom Small	69.82	7.15	102	Kerala
Cardamom Large	26.55	5.39	203	Sikkim
Ginger	67.20	233.66	3477	Kerala
Turmeric	124.60	487.40	1912	Andhra Pradesh
Coriander	521.60	308.10	590	Rajasthan
Cumin	288.83	116.26	402	Rajasthan
Fennel	27.439	368.96	1344	Gujarat
Fenugreek	33.590	314.13	945	Rajasthan
Chillies	831.500	821.80	988	Andhra Pradesh
Garlic	98.5	464.00	471	Madhya Pradesh

Spices are well known as appetizers and are considered essential in the culinary art all over the world. Some of the spices possess anti-oxidant properties and the others are used as preservatives in some foods. Many of them possess medicinal properties. Most of the spices are used in the industries in the preparation of varieties of food products as well as value added and consumer articles like cosmetics, tooth pastes, soaps etc. As India is the largest producer of spices, so is their consumption in the country. There is no Indian Cuisine without the addition of one or more spices. The kind of quality of spices used by the people vary from state to state and region to region. India holds first position in the per capita consumption of spices, which is around 2200 gm, while it is around 800 gm. in U.S.A., the largest importer of spices. However, as far as black pepper consumption is concerned, the per capita consumption in India is below 20 gm. as against 160 gm. in the U.S.A. It is needless to say, that, India will continue to hold the unique position in the consumption of spices so long as their uses are intermingled with dietary and social habits.

The consumption of spices varies from one country to another and is influenced to a large extent by the size of the population and the rate at which it grows. It is also influenced by the disposable income, which in the case of developing countries is a major factor. In the developed countries, spices are used in the industrial sector, principally in food processing whereas in developing countries, spices are mainly consumed in individual households. In general, the social habits, particularly those of cooking and eating, determine the overall per capita consumption of spices in both the developed and developing countries. These aspects decide the overall situation in the export performance of a country like India.

India exports entire gamut of spices. The major ones are pepper, chilli, turmeric, ginger dry, cardamom, cumin, coriander, fennel, fenugreek as well as spice oils and oleoresins. Malabar pepper, Cochin ginger, Allepey green cardamom, Allepey finger turmeric and Sannam chillies are some of India's favoured and preferred spice varieties in the international markets. In terms of volume, exports have recorded a whooping growth of 225 per cent from 70,279 tonnes in 1987-88 to 2,31,389 tonnes in 1998-99. The shift in emphasis from commodities in whole form to value added products has given a new dimension to the spice industry in the country. Exports of spice oils and oleoresins have recorded substantial growth from 1,672 tonnes valued at Rs. 86.76 crore in 1994-95 to as high as 2,750 tonnes worth rupees 300.77 crore in 1998-99. As a whole, India has a global monopoly in spices oils and oleoresins. On the other hand, India has not yet succeeded in certain spices like clove, cinnamon, nutmeg etc. and the requirements are met through imports. Indian spices and spice products are mostly exported to East, European countries, Central Asian countries, American zone and to the Middle East countries in large volumes and to the African and Australian zones in small volumes. As a whole, the recent performance of India is improving in it's exports of spices, still it has not yet recovered in the international market for spices. Once it had the monopoly in it, for this, there appears to be the prevalence of several hurdles in this industry.

These hurdles are lower productivity, non-availability of high yielding disease resistant quality planting materials, improper marketing activities, price fluctuations, increasing cost of production, increasing level of domestic consumption, competition from the new entries, stringent sanitary and Phyto sanitary conditions, demand for organic spices, pests and diseases etc. So as to regain the status in the international market and to improve Indian position both in terms of volume and value these problems have to be tackled with the help of a planned strategy. Again, there is also the need to improve the production of tree spices, vanilla etc., which are having the history of more than 150 years.

The world demand for spices and its products are ever increasing, further more, spices and herbs are building blocks to a series of value added derivaties—such as spice oils, oleoresins, food colours, mint oils, hydroxycitric acid, ground spices, curry powder, freeze dried green pepper, green and pink pepper in brine. We have the expertise and world class facilities to manufacture all these products which now dominate the international market. So the future prospects for the growth of Indian spice industry lies in focusing on value-added products, consistent drive on quality and ensuring competitiveness through enhanced productivity. Reduction in exportable surplus on account of increasing demand as can be observed from Table-II, vagaries of nature etc., are the major concerns of the industry. In this regard, the Spices Board and other organizations engaged in research, development and export promotion are also working positively over the years and are trying to maintain India's image in the international market as the only country in the world capable of supplying almost all spices.

TABLE-II : Production and Export of Spices in India (1997-98)

Spices	Production (in tonnes)	Export Volume (in tonnes)	Percentage of Exports in the total production
Black Pepper	57,270	36,026	62.9
Cardamom Small	7,150	297	4.15
Cardamom Large	5,390	1,703	31.6
Ginger	2,33,660	28,310	12.11
Turmeric	4,87,400	26,838	5.5
Coriander	3,08,100	20,901	6.8
Cumin	1,16,260	16,195	13.9
Fennel	36,896	12,159	32.9
Fenugreek	31,413	5,570	17.7
Chillies	8,21,800	42,172	5.1
Garlic	4,64,000	3,975	0.9

Eventhough, different spices are grown in India, some of the major and minor spices are taken into consideration here, for discussion, mainly on the basis of their economic importance in the economy. Efforts are made to record the available latest data in connection with area, production, productivity and for exports. The main intention of this volume is to highlight the importance of the spice, industry in Indian economy as they contribute a lot in the process of economic development through the much needed foreign exchange. An effort is made here, to highlight the major problems of this sector and an attempt is also made to suggest certain remedial measures to overcome these on the basis of personal experience as an agriculturist and by gathering information from various research findings which are published in various books and journals.

2

PEPPER

Black pepper or the "Golden Vine" is the most important of all the spices grown in India. It is also rightly called as the "King of Spices" or even as "Black Gold" of India because of its tremendous economic importance. Black pepper is the dried mature but unripe berries of Piper nigrum L, a branching vine or climbing perennial, is one of the most important and earliest known spice crop of India. It is a native to Western Ghats of India. In the early historic times, pepper was widely cultivated in the tropics of South-East Asia, Brazil, Indonesia, Malaysia, Sri Lanka, Vietnam and Cambodia, being the other pepper producing countries of the world. The world pepper production is estimated at 217, 164 tonnes for 2000, up from an estimated 212,200 tonnes for 1999. Substantial output gains are expected by 2003 as the planted areas for pepper are increasing.

Till the late 1940's, India's share in the world production was 80 per cent, and it came down to 66 per cent in 1950, this further declined to 26 per cent at per cent. On the otherhand, the shares of Malaysia, Vietnam, Brazil and Indonesia have steadily improved over the years.

Area, Production and Yield of Pepper In India

In India, pepper cultivation is mainly confined to Kerala, Karnataka, Tamil Nadu and Andaman and Nicobar Islands and to a limited extent in Assam, West Bengal and Andhra Pradesh. Kerala alone contributes above 95 per cent of the total production

TABLE-I: State-wise Area, Production and Yield of Pepper

(area '000 ha, production-'000 tonnes, yield-kg/ha)

State	1995-96			1996-97			1997-98		
	Area	Production	Yield	Area	Production	Yield	Area	Production	Yield
Karnataka	3.55	0.88	248	3.82	0.93	243	3.83	0.94	245
Kerala	190.84	59.94	314	172.60	53.77	311	173.86	55.52	319
Tamil Nadu	3.21	0.67	209	3.40	0.80	335	3.42	0.72	210
Andaman & Nicobar Islands	0.08	0.08	190	0.43	0.80	186	0.43	0.08	186
Pondichery	0.01	0.01	—	0.1	0.01	—	0.01	0.01	—
All India	198.03	61.58	311	180.26	55.59	308	181.55	57.27	315

Source: Department of Economic and Statistics, New Delhi.

of pepper. The total area under pepper in Kerala is about 174 thousand hectares and the production is around 56 thousand tonnes, while in Karnataka the area is about 3.83 thousand hectares and the production is 0.94 thousand tonnes as can be seen from Table-I, which gives the idea about the domination of Kerala in terms of area as well as production over the years. In terms of productivity, the all India average is 315 kg/ha. while that of Kerala is 319 kg./ha. and that of the other states, it is below average.

As far as total area and production of pepper in India over the years is concerned, it shows a positive trend upto 1995-96 since 1970-71 as shown in Table-II; however, since then it shows a negative trend both in terms of area as well as production. The maximum production of 61580 tonnes was observed in 1995-96 through an area of 198030 hectares. In 1997-98 the total area under pepper was 181550 hectares and the production was 57270 tonnes.

TABLE-II: Area and Production of Pepper in India

Year	Area (in hectares)	Production (in tonnes)
1970-71	119960	26160
1975-76	111930	25570
1980-81	109290	29490
1985-86	125120	34000
1989-90	155780	38790
1992-93	189390	50760
1993-94	190990	51320
1994-95	193270	51320
1994-95	193270	60740
1995-96	198030	61580
1996-97	180260	55590
1997-98	181550	57270

Source: Department of Economics and Statistic, New Delhi.

In Kerala, Idukki district covers an area of 43.40 thousand hectares under pepper and stands first in the state followed by Waynad, Kannur and Kollam districts. The cultivation of pepper is taking place mostly along with coconut, arecanut or as a single crop.

Varieties

There are many varieties of cultivated pepper. They differ mainly in the time taken to mature, the length of the spike and the size of the berries. Each tract has its own selection of popular varieties known by different names in different regions. The important varieties are Balankotta, Kalluvalli, Cheriakodi, Uthirankotta, Cheria Kanikadan, Chola, Perumkodi, Morata. The hybrid variety called Paniyur-1 has been evolved which combines all the desirable characteristics of Uthirankotta and Cheriakanikadam and it gives three to four times the yield of other varieties. The improved varieties which are recently made available are Paniyur-2, Paniyur-3, Paniyur-4, Paniyur-5, Subhakara, Sreekara, Panchami, Pournami and PLD-2. Among these Paniyur-2 contains highest piperine and the berries of Paniyur-1 and Paniyur-3 are very bold and these fetch premium price in the export market. Various kinds of pepper known to the spice trade are named after the districts in which they are produced or the neighbouring ports through which they are exported. Thus the pepper exported from Allepey and Tellicherry are generally known as 'Malabar Allepey' and 'Malabar Tellicherry' respectively.

The pepper vines commence bearing during the third year and it takes nearly 6-8 months from flowering to harvesting. The season for harvesting mature berries is November-February in the plains, and Jan.-March in the hills. Harvesting is generally done by plucking the spikes, then they are spread on the floor and after separating the berries, they are dried for 4-7 days in the sun and this is called as black pepper.

Uses

Pepper is used for a variety of purposes. The ancient Aryans considered it as a powerful remedy for various disorders of the anatomical system and prescribed it as an effective cure for dyspepsia, malaria etc. It is largely used by meat packers and in pickling, baking, confectionery and preparation of beverages. One of the principal values of pepper is it's ability to correct the seasoning of dishes.

Oil of pepper is a valuable adjunct in the flavouring of sausages, canned meats, soups, table sauces and certain beverages and liquor. It is used in perfumery, particularly in bouquets of the oriental type to which it imports spicy notes different to identify.

Pepper for the Market

Various products of pepper are having good demand both in the internal and external markets. The products are popularly known as the value-added pepper products. These value added pepper products may be classified into three main groups via. (1) Green pepper based products; (2) Black pepper based products and (3) Pepper bye-products.

(1) Green pepper based products

(a) Canned and bottled tender green pepper

For preparing this product, four to five months old immatured green pepper berries are harvested and then despiked and immediately steeped in chlorinated water for about 30 minutes and then washed thoroughly under running water. Normally acetic acid/citric acid is used in the preservative solution which are 2 per cent brine solution in case of canned pepper and 15 to 20 per cent in case of bottled green pepper. It is reported that Madagascar is the leading producer of canned green pepper.

Canned green pepper is in great demand in many European countries like Federal Republic of Germany, France, Belgium, Finland, Denmark along with USA, Japan and the Middle East countries. It is preferred to black pepper due to the freshness, natural attractive green colour and exotic aroma. India and Madagascar are the leading exporters of this product.

(b) Dehydrated green pepper

In the bottled/canned green pepper in brine the quality is not uniform and if the green pepper used is too immature, the cooked products become mushy sand if it is over matured, then it is hard to bite. The efforts of CFTRI, is responsible for the development of a new product called "Dehydrated Green Pepper" which is a better substitute for the canned or bottled green pepper. The intention of this was to reduce the cost of packaging and transportation and to get uniform green coloured product. For preparing this product, it requires freshly harvested, (20-30 days before ripening) green pepper, which is de-stalked, freed from pinheads, light berries etc. and washed. The berries are then subjected to heat treatment by dipping them in hot water for about 20-25 minutes, then the material is rapidly dried in hot air at a medium temperature to preserve the green colour. The well dried produce is thereafter-cleaned, graded and packed in polythene lined bags or teachests. This product remains green over long periods of storage and reconstitutes in original shape and gets fresh appearance when soaked in hot water. As this product possesses a very fresh smell of harvested green pepper and higher oleoresin content it has found new avenues besides garnishing of steaks. The entire volume of this product is exported from India. The major importers of this are Federal Republic of Germany and Belgium. The other consumers are Netherlands, Switzerland, Sweden, Japan, USA etc.

(c) *Frozen green pepper*

It is a new product prepared for the diversification in pepper exports. This is in demand in German FR, France etc. This product is considered as superior to green pepper in brine or dehydrated green pepper because of its better flavour, colour, texture and natural appearance and the cost of packaging is also lower for this. Frozen green pepper is gaining popularity nowadays because of its superiority in every respect.

(d) *Green pepper pickle*

It is very much popular in many states like Kerala, Karnataka, Tamil Nadu, Gujarat, Maharashtra etc. People relish it with rice meals as an appetizer. By mixing it with shredded fresh ginger, it gives more taste and piquant.

Green pepper berries are also mixed with lime pickles, mango pickles etc. along with sliced fresh ginger. They are popular all over India.

Green pepper is also used in soups, rasams etc. The berries give the exotic taste to westerners while eating it in conjunction with stalks and other meat products. Green pepper is also used in garnishing of salads and other foods.

(e) *White pepper*

It is prepared by removing the outer pericap or skin of the harvested berries by any one of the following methods:

(1) *Water steeping technique:* (*a*) using ripe fresh berries:

This is a traditional method in which the harvested ripe spikes or berries are packed in gunny bags or steeped as such in water tanks or running water for 7-10 days. The pin heads and light berries which float are separated. Once the skin is softened they are rubbed by hand or trampled, these berries are then sun dried and sold as white pepper.

(2) *Steaming or boiling technique*

This process consists of steaming or boiling fresh matured berries for about 15 minutes. The boiled or softened berries are de-skinned, washed, bleached and sun dried.

(3) *Decortication technique*

White pepper is also prepared by decorticating black pepper in decorticating machines. However, this incur loss of berries to about 20-25 per cent due to breakage.

In India, white pepper has not been made on commercial scale. Only small quantities are produced in certain pockets and exported. The reasons for this are first of all, the growers are not willing to allow the berries on the vines for ripening because of fear of loss due to shedding or picking up by birds. Secondly, the farmers are not interested in converting black pepper into white pepper, since it is not an attractive proposition for them.

(2) **Black Pepper based Products**

(*a*) *Black pepper powder*

In India, pepper powder for export is prepared from good quality garbled black pepper, mostly MG I grade. A few organized units in India are manufacturing pepper powder in large quantities both for internal and external markets. For this product there is a huge demand in Russia, Yugoslavia, Saudi Arabia, Kuwait, UAE, UK, USA etc.

(*b*) *Pepper oil and oleoresin*

India started producing spice oils and oleoresins on a commercial scale towards the end of 1960s and since then it has gathered momentum.

The flavour of pepper is on account of the volatile oil. Pepper oil is obtained by the steam distillation of ground pepper and it's yield is 2-4 per cent of the raw material. The composition of pepper oil varies widely depending on the nature of the raw material used i.e., pepper varieties, grades, storage conditions, processing methods etc.

Pepper oleoresin is prepared by solvent extraction of ground pepper. Solvents used are generally acetone or chlorinated solvents like methylene chloride or ethylene dichloride. Pepper oleoresin contains the oil responsible for the aroma and the resinous portion containing the pungency factors.

Pepper oil and oleoresins are used in pharmaceutical, perfumery and for flavouring canned meats, sauces and soups etc. Pepper oleoresin has become more popular since it has advantages of flavour strength, easy disperson, long shell life, hygienic etc. The export of pepper oil and oleoresin commenced in the early seventies and has grown steadily over the years.

(3) *Pepper Bye-products*

Three different bye-products are available in the markets viz., the pepper rejections or waste, varagu or the unfertilized buds and the stems and inflorescence stalks. In order to economise the use of pepper as a condiment and to replace it in times of scarcity, many products having the characteristic taste and pungency of pepper has been prepared by patented processes, particularly in the USA. Underdeveloped pepper berries of low specific gravity but rich in oleoresin, have been extracted with solvents for further processing into spice paste. A homogenous solution of the oleoresin is being prepared for the use in salad dressing.

Apart from the above said bye-products some other bye-products like pep-sal, pepper husk, light pepper, pepper pinheads etc. are also prepared and marketed both in the domestic and external markets.

Marketing of Pepper

In India, marketing of pepper is mainly in the hands of a few traders and exporters. In the normal case there is no Government intervention in the marketing procedure or in the maintenance of price. Pepper marketing is traditionally a private business. There are a number of intermediaries like itinerant/village merchants, wholesalers, commission agents and exporters who are responsible for the disparity between the price paid by the consumer and the price received by the producer. The growers in the village get only 60-70 per cent of the retail price in the distributing markets.

Normally, the farmers dispose pepper either locally to a village merchant or shop, a cooperative or even to a terminal market. The choice depends on a number of factors such as loyalty or financial obligations to particular traders, the distance from terminal markets, the prevailing market price etc. Selling the standing crop to pre-harvest contractors is also in-vogue in some areas in which case the contractors make advance payment to the farmers. Collection centres operate within short distances and also in every town. Here the merchants purchase the produce and after accumulating sufficient stock or when the terminal market price is sufficiently attractive, the commodity is bagged for transport to the terminal markets. The main terminal or wholesale marketing centres for pepper in India are Calicut, Tellichery, Cochin and Allepey. In Kerala, Sirsi and Mangalore in Karnataka, Mumbai in Maharashtra, Chennai in Tamil Nadu and Delhi. Cochin is the main market centre from where the exports are taking place in large volume.

In these terminal markets pepper brought from merchants are received by commission agents and it is then traded between brokers or dealers. Exporters purchase from the commission agents or dealers. The exporters store the pepper in their godowns until an export order is received. Then he grades it, obtains the grade certificate and finally consigns the product to the shipping agents.

There are a few regulated markets for pepper in Karnataka, while in Kerala, pepper has not been brought under regulation so far. Although there are a few co-operative marketing societies in Kerala handling pepper, but they are not effective because of poor financial resources. Apart from these, there is an Apex body of primary co-operative marketing societies "Kerala State Co-operative Marketing Federation", both of these are transacting in spices, in particular pepper. This Federation is not only processing pepper through co-operatives but also exports it.

Grading of Pepper

The price of black pepper is decided according to the general appearance of the produce, percentage of moisture, presence of light berries, pinheads etc. The pepper brought by the growers or village merchants for sale are mostly in the ungraded or semigraded form. Garbling and grading is normally resorted to either the wholesalers level or at the exporters level. A few merchants in the assembling markets, however, resort to garbling and grading before disposing it off to the wholesalers or exporters. The wholesalers or exporters dry the produce again after washing, winnow it to remove chaffs, light berries and pinheads and sieve it with sieves of different mesh sizes to get berries of different grades.

In order to ensure quality of the pepper exports from India it is subjected to compulsory quality control and inspection. Grading of pepper under AGMARK for export was therefore, made compulsory since Jan. 1963. The Agmark grade specifications prescribed under the pepper grading and marketing rules 1961 (which was later revised in 1969). For the export of pepper under different trade descriptions took into consideration different quality factors such as garbled and ungarbled berries, moisture content and other aspects such as colour, surface shape, freedom from moulds or insect or any other adulterant etc. Specifications have been made separately for different types of pepper such as "Malabar garbled/ungarbled", "Tellichery Garbled", "Garbled Light", "Pinheads" etc. On the whole, there are about nineteen special grades of black pepper grading under Agmark for export which is mostly done in the wholesale marketing centres. Among the different grades of pepper that are exported, Malabar Garbled black pepper (MG I) is the most popular grade which accounts for about 80-85 per cent of the overall exports. Besides, pepper in the whole form, pepper powder is also graded under Agmark for export. However, black pepper is not graded under Agmark for the internal market.

Price Behaviour

As the production of pepper is seasonal and unlike other agricultural commodities, its price is largely influenced by the external demand from other countries. Vagaries in supply and elements of speculation are the twin problems connected with the world price of pepper. In the markets, the producers are paid more than the economic price when the supply is low and it reverses when the supply is supposed to be slightly higher than the demand, resulting in a greater instability in price. Generally, the prices of pepper in the market are depressed in the months of Jan.-Feb. mainly due to fresh arrivals in the market and it registers an increase from July-Aug. Following a fall in arrivals and later it goes on increasing and stay at a higher level with frequent fluctuations depending on foreign demand till the arrival of the next season. Again, the price of pepper varies from one market to another depending upon the demand and supply and also costs involved in its marketing activities. The ata given in Table-III indicates the same thing where average annual wholesale prices of pepper in important markets in India are shown for the period 1994-1999.

Export

Black pepper is rightly considered the 'King of Spices' as judged from the volume of international trade, being the highest among all the spices known. Apart from this, pepper is one among the major items of our exports which fetches the major part of our export earnings. India exports black pepper, white pepper and pepper products like dehydrated green pepper, pepper oil and pepper oleoresin to USA, Russia, Canada and East European countries in large quantities. As far as the export of black pepper is concerned in 1997-98, USA alone imported 16,447 tonnes worth Rupees 22,60,719 thousands followed by Russia, Canada and Italy as can be observed from Table-IV.

In terms of total volume of black pepper exports are concerned, it was 15,641 tonnes in 1950-51 and in 1960-6l, it was 17,202 tonnes and the maximum volume of 48,743 tonnes was exported during

TABLE-III: Average Annual Wholesale Prices of Pepper
in Important Markets in India (Rs./Quintal)

Year	Kerala					Karnataka	Maharashtra	Tamil Nadu	Delhi
	Calicut		Tellichery	Cochin	Allepey	Sirsi	Mumbai (Malabar)	Chennai Attom	Delhi
	Nandan	Wynaadan	Chettan	Garbled	Palai				
1994-95	6020	6361	5886	6392	6080	5828	5584	6793	6683
1995-96	7382	7889	7271	7810	7704	7567	7549	8279	8687
1996-97	7750	8283	7669	8360	NT	7748	8863	9036	9087
1997-98	15961	17171	15918	17700	17568	16215	12229	17532	17733
1998-99	19174	20629	19378	20273	20118	19781	—	20246	22414

Source: Price Inspector, Market Intelligence, Allepey, India Pepper & Spices Trade Association, Cochin
Agricultural Produce Market Committee, Sirsi,

Commissioner of Statistics, Madras,

D of D Marketing (S & R) Bombay

Regional Statistical Officer (MT) Calicut/Tellichery.

TABLE-IV: Country-Wise Export of Pepper from India

(Qty.—tones, Value '000 Rs.)

Country	1995-96		1996-97		1997-98	
	Qty.	Value	Qty.	Value	Qty.	Value
Russia	3318	27955	3335	298654	3149	451492
USA	7740	540873	25427	2193858	16447	2260719
Canada	1479	115912	2360	190693	1864	252270
Italy	2172	174500	1898	162851	1816	287211
Czechoslovakia	105	8822	154	1240	—	—
Yugoslavia	—	—	15	1333	—	—
Poland	1290	103849	882	77215	961	107341
Rumania	146	10258	—	—	—	—
Saudi Arabia	283	22941	216	17001	79	5781
Total (including others)	26244	1962986	47893	4123182	36026	4360457

Source: DG of CI and S, Calcutta.

the period 1993-94. The figures for pepper exports from India are given in Table-V, which shows that there are ups and down in the exports. In 1999, pepper exports from India rose 27 per cent more from the previous year. During the period 1998-99, pepper exports from India were 33,000 tonnes and it is expected to move upto 42,000 tonnes in 1999-2000.

TABLE-V: Export or Pepper From India

Year	Quantity (M.T)	Value (Rs. '000)
1950-51	15641	204033
1955-56	13336	47083
1960-61	17202	84966
1965-66	26305	111022
1970-71	17970	152485
1975-76	24226	338837
1980- 81	26364	389487
1985-86	37620	1724849
1990-91	31871	1110604
1992-93	22684	751126
1993-94	48743	1890968
1994-95	37264	2366422
1995- 96	26244	1962986
1996-97	47893	4123182
1997-98	36026	4360457
1998-99	35100	6350000
1999-2000*	40000	8000000

*Estimated.

Source: DG of CI and S, Calcutta.

India's share in the international market of pepper was 75 per cent in 1947-48 but it declined to 18.5 per cent in 1983-84 and a further decline was observed in 1984-85 to 16 per cent and at present it is below 25 per cent. On the other hand, the shares of Malaysia, Vietnam, Brazil and Indonesia are increasing over the years.

World pepper export which showed a healthy growth of 10 per cent as per the provisional estimates for 1999, is forecast to grow in 2000 by about 4 per cent according to International Pepper

Community. According to it, the aggregate export of pepper during the year 1999 at 1,51,830 tonnes was up from 1,38,150 tonnes in the previous year (Table-VI).

TABLE-VI: Pepper Exports from Major Pepper Producing Countries

(*in Tonnes*)

Country	1998	1999	2000*
Brazil	17,250	18,000	20,800
India	33,000	42,000	36,000
Indonesia	37,735	33,800	34,000
Malaysia	17,830	21,000	22,000
Vietnam	22,000	28,000	32,000
Others	10,335	9,030	13,000
TOTAL	1,38,150	1,15,830	1,57,800

*Projected

Source: Jakarta based IPC.

India exports both pepper oleoresins and pepper oil in large volume to USA, UK. France, Canada and Belgium etc. As far as pepper oleoresins export is concerned, USA is the major buyer which purchased 2,34,472 kgs. worth Rupees 2,24,533 thousand in 1997- 98 followed by UK. during the same period. The total export of oleoresin in 1997-98 was 5,61,845 kgs. valued at Rupees 5,56,601 thousand and that trend over the years shows the positive aspects. (Table-VII). For pepper oil German FR is the main market, which purchased 15140 tonnes of it in 1997-98 and USA's import was 9,102 tonnes during the same period. In terms of total volume of pepper oil exports, it was 36,162 tonnes in 1997-98 worth rupees 41,993. However, there appeared slight variations in the volume of exports during past years as can be seen from Table-VIII.

The above aspects in connection with area, production, productivity and India's exports of pepper clearly shows that the country has lost everything in this, as far as the world production and markets are concerned. This type of situation appears to be there because of the prevailing problems of the sector. These are:

TABLE-VII: Country-Wise Export of Pepper Oleoresins from India

(Qty.: kg., Value: '000 Rs.)

Country	1995-96		1996-97		1997-98	
	Qty.	*Value*	*Qty.*	*Value*	*Qty.*	*Value*
Belgium	1500	969	5002	3265	1592	1965
Canada	21970	12315	18248	10887	30625	29896
France	27190	18156	32558	19118	39854	41823
Japan	7220	4855	3814	2651	7355	7735
UK	85999	49929	90639	54779	80024	73094
USA	257382	165410	245066	151321	234472	224533
Total (including Others)	559514	350120	518546	324099	561845	556601

Source: DG of CI & S, Calcutta.

TABLE-VIII: Country-wise Export of Pepper Oil from India

(Qty. in tonnes, Value '000 Rs.)

Country	1994-95		1995-96		1996-97	
	Qty.	Value	Qty.	Value	Qty.	Value
German FR	9541	8543	9823	10955	15140	17017
Netherlands	1525	1105	1192	2004	500	512
UK	8865	5832	3918	3171	3480	3528
USA	11755	9027	12058	11654	9102	9725
Total (including others)	38164	30547	34732	35888	36162	41493

Source: DG & CI & S, Calcutta.

1. The yield of pepper in India is very low as compared to other pepper growing countries. The reasons for this are: (a) Pepper cultivation is mostly in the hands of small holders, which indicates that there is virtually no organised cultivation on a plantation scale; (b) Poor management; (c) Cultivation of poor genetic material; (d) predominance of old and unproductive vines and poor yielding; (e) pests and diseases like minor ones, pepper mealy bug, pepper scales; (f) crop losses due to foot-rot or quick wilt diseases, caused by the soil borne fungus Phytophth or a capsici.

2. Non-availability of credit, subsidy facilities reduced the interest of the farmers to extend the area under pepper.

3. The cost of production is high in India. This is because of the non-availability of low cost technology in different production aspects.

4. Non-availability of well rooted cuttings of high yielding varieties.

5. Lack of enough programmes for large scale production.

6. Motivation of farmers to adopt the new technology through field demonstration has not been carried out sufficiently.

7. High cost of inputs and lack of enough incentives also hinder the crop development.

8. In the normal case, there is no government intervention in the marketing procedure or in the maintenance of price.

9. In the international market, lack of quality control methods reduced the scope for expanding the exports. The traders adulterate pepper berries with papaya seeds, dust, dirt, stems etc.

10. Excess purchase of premature pepper by the oleoresin industry adversely affects the volume of exports.

11. Competition from Vietnam, Brazil, Indonesia in the international market.

Prospects

The world demand for pepper has been increasing in recent years and prices are attractive hence there is a need to build this sector in a sound manner. Again the popularity of seasoned and hot foods is increasing in their existing markets, so the industry would need to prepare for higher future output by promoting the unconventional usage and applications of pepper. World consumption of pepper is growing annually at an average of 2-3 per cent; Vietnam's production may exceed 40,000 tonnes in 2005 and 100,000 tonnes in 2010 and smaller producers like Malaysia and Sri Lanka are also making efforts to raise pepper output. To counter the situation, the country has to take up steps to promote pepper production by overcoming the above said problems. In this regard, a proper long-term planned strategy is needed and it has to prescribe the required solutions like:

1. To increase the production and productivity, there is the need to replace the senile, diseased and unproductive vines by high-yielding disease resistant varieties. Large scale production and distribution of rooted cutting of high yielding varieties through rapid multiplication methods are required. Again there is the need to identify the varieties which have high production potential and better export demand.

2. Need to rejuvenate unproductive gardens by replanting and adopting scientific cultivation methods.

3. Need to motivate the farmers to follow improved cultivation methods. Along with this, there is also the need to educate the farmers to adopt the latest technology for harvesting, processing, storage and marketing.

4. Government agencies, co-operatives and voluntary organizations should develop the common facilities for quality improvement, storage and marketing.

5. Need to expand the area under pepper with the help of a package programme.

6. Research stations should supply the necessary pre-and post-harvest information to the farmers.

7. Need to produce and market new pepper products like dehydrated pepper, freeze dried green pepper etc.

8. All those research activities which are undertaken in the labs in connection with the uses of pepper, has to be converted into commercial products.

If all these steps are taken in spirited manner with efficiency, then there is a vast scope for Indian pepper in the long run in the world market.

3

CARDAMOM

Cardamoms are the dried capsules of a small group of species or plants belonging to the family Zingiberaçeae, which contain seeds possessing a pleasant characteristic aroma. There are three distinct varieties viz. Malabar type, Mysore type and long Ceylon wild type. The first two are commercially far more important than the third one. The Malabar and Mysore types are popularly known as small cardamom or Chhota Elachi which actually constitutes the second most important 'national spice' of India and is rightly known as the 'Queen of Spices'.

Area and Production of Cardamom in India

In India, cardamom is grown in the Western Ghats of Southern parts and in the North Eastern Himalayan states like Sikkim, Meghalaya, Assam, Tripura, Arunachal Pradesh and West Bengal. Small cardamom is mainly grown in the southern states viz. Kerala, Karnataka and Tamil Nadu while in the other states the large cardamom is grown. The total area under small cardamom in 1977-78 was around 91,480 hectares and the production was about 3,900 tonnes, the area under this has increased to 105,000 hectares in 1987-88 but the production declined to 3,200 tonnes whereas in 1997-98, the area went downwards to 70,000 hectares and the production went up to 7,150 tonnes. (Table-I, II).

Among the states, where small cardamom is grown, 61 per cent of the total area is in Kerala, 31 per cent in Karnataka and 8 per cent in Tamil Nadu and in terms of yield per hectare in Kerala, it is 126 kgs/ha. and in Tamil Nadu it is 90 kgs/ha while in Karnataka it is only 58 kgs/ha.

TABLE-I: Area and Production of Small Cardamom in India

Year	Area (in hectares)	Production (in tonnes)
1970-71	91480	3170
1975-76	91480	3000
1980-81	93950	4400
1985-86	100000	4700
1989-90	81000	3100
1992-93	70530	4380
1993-94	74790	6440
1994-95	74000	6420
1995-96	75120	7380
1996-97	72520	7290
1997-98	69820	7150

Source: D of E and S, New Delhi.

As far as large cardamom is concerned, the total areas under it in India over the years has been around 27 thousand of hectares. The production in 1980-81 was 4000 tonnes, in 1989-90 it came down to 3260 tonnes and in 1997-98 it was 5390 tonnes. In terms of production, it has been increasing since 1995-96. (Table-III)

India's contribution in the world production has been declining over the years and at present it is around 45 per cent of the total, while that of Gautemala which was around 16 per cent in the 70's, has increased to 46 per cent at present.

Uses

A good portion of the cardamom produced in India is consumed internally, for chewing or as a matsicatory, as a common ingredient of special seasonings and curry powders and for flavouring sweetmeats, pastries, cakes, other bakery products, ginger bread, puddings, kheer and culinary preparations. It is also used in sweat pickles. Its essential oil is used for flavouring certain bitters and liquors and also in the manufacture of perfumes. In Arab countries, a special cardamom flavoured coffee is an attraction in the social and religious functions.

TABLE-II: State-wise Area, Production and Yield of Small Cardamom in India

(Area '000 ha, Production '000 tonnes)

State	1995-96			1996-97			1997-98		
	Area	Production	Yield	Area	Production	Yield	Area	Production	Yield
Kerala	44.25	5.38	122	43.05	5.40	125	43.05	5.43	126
Karnataka	25.57	1.50	58	24.07	1.39	58	21.41	1.24	58
Tamil Nadu	5.30	0.50	94	5.40	0.50	93	5.36	0.48	90
Total	75.12	7.38	90	72.52	7.29	100	69.82	7.15	102

Source: D of E and S, New Delhi.

TABLE-III: Area and Production of Large Cardamom in India

Year	Area (in ha.)	Production (in tonnes)
1980-81	25685	4000
1985-86	23110	3380
1989-90	25690	3260
1992-93	26150	4830
1994-95	26270	4240
1995-96	26370	4330
1996- 97	26470	5320
1997-98	26550	5390

Source: D of E and S, New Delhi.

Cardamom has certain medicinal values too. Tinctures of cardamom are also made and used chiefly in medicines for windiness or stomachic. Powdered cardamom seeds mixed with ginger, cloves and caraway is helpful in combating digestive ailments. In medicine, it is used as a powerful aromatic stimulant, carminative, stomachic and diuretic. It also checks nausea and vomiting and is reported to be a cardiac stimulant also. Powdered seeds of cardamom boiled with tea-water imparts a very pleasant aroma to the tea, and the same can be used as a medicine for scanty urination, diarrhoea, dysentery, palpitation of the heart, exhaustion due to over work, depression etc. Eating a cardamom once daily with a tablespoon of honey improves eye-sight, strengthens the nervous system and keeps one healthy, however, excessive use of cardamom causes impotency.

In India, the total consumption of cardamom is the highest one in the urban areas; for example in Maharashtra, Goa, Bihar, Tamil Nadu, Punjab, Gujarat, Delhi, U.P., Karnataka etc. it is above 80 per cent of the total sale of cardamom in these states, while in Kerala, the rural consumption is the maximum one which is above 65 per cent as per the survey conducted by Spice Board.

The survey further reports that in Gujarat consumers prefer green colour, since, it is fresh. Thirumalathirupathi Devasthanam is one of the largest domestic consuming centres in India. As far as the market for cardamom in general is concerned, the retail consumer demand for it is as high as 71 per cent and the industrial demand is about 21 per cent. In terms of per capita consumption by the households, it is the highest one in North India with 80 gms and in South, it is 60 gms, in the West 59 gms. and in the East 43 gms. while the all India average is 60 gms. On the other hand as far as industrial use of cardamom is concerned, maximum percentage of 57 per cent in the total market is going towards Pharma Ayurvedic and for cosmetics it is 33 per cent. In terms of institutional market, pan beedi shops are the main markets for cardamom in the country. The Spices Board, sponsored market survey of small cardamom, 1997 reveals that the domestic market for Indian cardamom in 1996-97 was 6387 tonnes which was 7400 tonnes in 1995-96.

Marketing

Cardamom, for the market is obtained from the production and despatching centres like Bodinayakanur, Virudhunagar, Sakaleshpur, Mercara, Cumbum etc. Cumbum supplies the bleached white cardamons while the other centres supply green cardamons. The marketing channel consists of the wholesalers, retails and the panwalas. As far as regulation is concerned, it is mainly to ensure a fair price and the timely payment. As a whole, the market structure of cardamom in India consists of a movement:

(1) Growers — Auction centres — Line Traders — Retailers Institutions and Households.

(2) Growers — Petty merchants — Commission Agents — Wholesalers — Retailers.

(3) Growers — Petty merchants — Dealers — Wholesalers — Exporters.

(4) Growers — Petty merchants — Commission Agents — Wholesalers — Brokers — Exporters.

(5) Growers — Petty merchants — Commission Agents — Wholesalers — Brokers — Retailers — Consumers.

(6) Producer — Exporter.

However, the entry of dealers and exporters to the market is regulated by the Cardamom licensing rules of 1977. In general, the marketing of cardamom is undertaken by the Cardamom Marketing Corporations auction centre.

The important internal markets for cardamom in India are Delhi, Kanpur, Lucknow, Allahabad, Agra, Bombay, Nagpur, Calcutta, Ahmedabad, Ajmer, Jaipur, Amritsar, Ludhiana, Patna, Cuttack, Bhopal, Indore, Hyderabad, Vijayawada and Rahajmundry. Bombay is the major export market in India.

Depending upon the variety of Cardamom, the price in market is quoted by the traders. The main varieties for the trade include the superior variety 8 mm; bold fast green fetches the maximum price but the demand for it is less whereas the demand for inferior quality is the maximum one. Apart from these classifications, the other prevailing varieties are AGEB, AGEB fast colour, Standard quality AGEB, AGB superior, AGS and AGSI. Among these AGEB fetches highest price followed by the other varieties respectively.

Exports

Till 1985-86, India was the main producer as well as the exporter of small cardamom in the global market, but the entry of Gautemala in 1985-86 relegated India to the second position in global trade. The share of India in the global market in 1970-71 was 55 per cent of the total, while that of Gautemala was 32 per cent and at present the reverse position can be noted.

In the international market, small cardamom has got an extensive demand and for large, the demand is small. Among the major spices exported from India to the global market, cardamom places second in terms of earnings. India exports cardamom, cardamom oil to Middle East countries, USA and East European countries.

In the year 1967-68, the total volume of exports of small cardamom from India was 1451 M. Tonnes valued at Rs. 70,261 thousands which went up to 2763 M. tonnes in 1977-78 and in terms of value it was Rs. 4,84,363 thousand, however it came down

sharply to 270 M. Tonnes in 1987-88 valued at Rs. 34,003 thousands
and an upward movement started since 1993-94 and reached the
maximum of 500 M. Tonnes in 1995-96 valued at Rs. 1,23,955
thousands, but once again it went downwards to 297 M. tonnes in
1997-98 valued at Rs.1,06,371 thousands (Table-IV). On the other
hand, the exports of large cardamom reached the maximum of 1784
M. Tonnes in 1995-96 valued at Rs. 1,23,495 thousands and it come
down to 1703 M. Tonnes in 1997-98 valued at Rs.1,26,005 thousands
(Table-V). As a whole, these figures show that there are ups and
downs in the export of cardamom from India over the years. This
type of situation appears to be there mainly because of the
problems of this sector. They are:

TABLE-IV: Export of Small Cardamom from India

Year	Quantity M.T.	Value Rs. '000
1964-65	1503	27177
1967-68	1451	70261
1970-71	1705	112160
1974-75	1626	133232
1977-78	2763	484363
1980-81	2345	347539
1984- 85	2383	648053
1987-88	270	34003
1990-91	379	102224
1992-93	190	75057
1993-94	387	145483
1994-95	257	76261
1995- 96	500	123955
1996-97	226	85967
1997-98	297	106371
1998-99	319	159704
1999-2000*	400	200000

*Target

Source: Directorate General of Commercial Intelligence and
Statistics, Calcutta.

TABLE-V: Export of Large Cardamom from India

Year	Quantity M.T.	Value (Rs. '000)
1964-65	257	1181
1967-68	91	960
1970-71	60	1190
1974-75	70	949
1977-78	218	4269
1980-81	225	5313
1984- 85	265	11653
1987-88	155	7022
1990-91	961	43155
1992-93	1225	88166
1993-94	1797	125696
1994-95	1293	81274
1995- 96	1784	123495
1996-97	1628	120953
1997-98	1703	126005
1998-99	1442	137986
1999-2000*	1200	110000

*Target

Source: D G CI and S, Calcutta.

Problems

(1) Domestic Problems

The production of cardamom has been fluctuating in India over the years mainly because of non-availability of quality seeds, lower productivity, credit facilities to the producers, irrigational facilities. Apart from these, a rise in the price of coffee has minimised the interest in its cultivation in Karnataka. Major cardamom growers are also slowly switching over to coffee and pepper adaptable to the high attitude as cardamom is becoming economically inviable. Again uncertainty in the prices of cardamom is compelling them to give up cardamom cultivation for ever. The input cost is higher in cardamom; besides, the crop requires frequent irrigation and care. For the small growers, who had taken financial assistance from banks, repayment was becoming a major hurdle. Involvement of smugglers because of the hike in sales tax, is a threat to the Government's revenue.

(2) *External*

The Indian cardamom has been largely purchased by the Middle Eastern market for a long period, but because of the arrival of Gautemala in a big way, the demand for Indian Cardamom declined in recent years. Gautemala supplies cardamom at a lower price than that of India, so it actually minimised the scope for Indian exports.

In terms of average productivity in India, it is around 100 kg/hectare while in Gautemala, it is around 300 kg/hectare. The total production in Gautemala has risen sharply over the years and at present it is above 16,000 tonnes, while in India it is not so. Apart from this there is lack of attention towards quality control limits in the export of cardamom from India which proves a passive and underdeveloped state of marketing.

Prospects

So as to promote the exports, first of all, there is the need to increase the productivity and production domestically which calls for long-term strategy of crop development. To achieve this distribution of better planting materials, better schemes for irrigating the plantations and other crop development methods have to be adopted. The improved varieties like Mudigere 1, 2, PV-1, CCS-1, ICRI-1, 2, 3, 4 should be made available to the farmers so as to increase the productivity. There is the need to educate the planters regarding the techniques of modern methods of cultivation. Apart from these, new areas which are eco-friendly to the use of high yielding varieties, should be brought under cardamom cultivation.

Along with these, the Spices Board should also draw up some new schemes to help the adoption of high productive techniques in the production of cardamom and it has to encourage the farmers to resort to mixed cropping.

Again, for export competitiveness, there is the need to reduce the price by a reduction in the cost of production. There is also the need to undertake publicity campaigns, participation in international fairs and exhibitions abroad. So as to maintain the improved quality, cardamom in consumer packs will be useful, for which there is a need to improve the post-harvesting methods. Hence, efforts are needed to increase the productivity, production, quality and expand the exports, then only India will be the 'Queen' in "Queen of Spices". Apart from these, an increase in the production in India can meet the domestic as well as external demand, since the internal market for cardamom is also increasing over these years.

4

CHILLIES

Chillies are the dried ripe fruits of the spices of genus capsicum. They are also called red peppers or capsicums and they constitute an important, well-known commercial crop used both as a condiment or culinary supplement and as a vegetable. Chilli was not known to Indians about 400 years ago, since this crop was first introduced into India by the Portuguese towards the end of the 15th century. Its cultivation became popular in the 17th century. Chilli is actually reported to be native to South America and its cultivation was known to the natives of Peru since prehistoric times.

Varieties

The varieties under cultivation differ in the size, shape, colour and pungency of the fruits.

The varieties of chillies are broadly divided into two groups namely (i) The long pungent type, including pickling type, used as a spice and (ii) the bell-shaped, non-pungent or mild, and thick fleshed type, popularly known as 'Simla Mirch' which is commonly used as a curried vegetable. 'Paprika' also belongs to the mild group. Most of the bigger red-coloured fruits cultivated and marketed the world over, including the chillies, paprica and capsicum, belong to the species Capsicum annum; the highly pungent small variety belongs to the species Capsicum frutescens. The early botanists selected the term Capsicum to designate the genus belonging to the family Solanaceae.

The term 'Paprika' is generally used for non-pungent (sweet) red capsicum powder. Red chile (pungent) and paprika are dehydrated and sold as whole fruits or ground into powder. In the United States, paprika is made from the New Maxican type chilli, whereas in Europe, Paprika is made from two main fruit types (*i*) A round fruit about the size of a peach called Spanish or Moroccan paprika. In fact, the Hungarian word for plants in the genus Capsicum is "Paprika" and it may be pungent or non-pungent.

The perennial are known as 'bird chillies' belong to C. frutscens Linn. The bird chillies are very pungent, short-lived and grow for 2 or 3 years. They are enlisted in the British Pharmacopoeia and find maximum use in pharmaceuticals.

The different varieties available for cultivation are 'NP 41' a high yielding pungent chilli and 'NP 46', another pungent chilli resistant to thrips. 'Hybrid 5-1-5' is high yielding and suitable for the production of green chillies. Among the non-pungent vegetable types, two American varieties, 'World Beater' and 'Bell pepper' and one Russian variety. 'RH 49' are high yielding. In Andhra Pradesh, the improved varieties 'Gl, G2, G3, G4 and G5' and four cultures, 'X200', 'Ca960', 'K196' and 'X197' are high yielding and are fast spreading in the state. The variety 'Gl' is high yielding and tolerant to thrips, has a persistent Calyx and is highly suited for export.

Area, Production and Yield of Chillies in India

In India, chilli is grown in almost all states. Andhra Pradesh has the highest area and production followed by Karnataka, Maharashtra, Punjab, Rajasthan, West Bengal and others. Productivity of chilli is the highest in Andhra Pradesh followed by Punjab, Rajasthan and Karnataka as can be seen from Table-I. The All India area under this crop during 1997-98 was at 831.5 thousand hectares and the production was at 821.8 thousand tonnes which was less than the area in 1996-97 that of 944.2 thousand hectares and in terms of production, it was 1066.4 thousand tonnes. The fall in the production as well as area under chillies was reported in states like Andhra Pradesh, Karnataka, Rajasthan and Tamil Nadu during the period. Table-II gives data for the trend in area under production of chillies in India over the years.

TABLE-I: State-Wise Area, Production and Yield of Chillies

(Area—'000 ha, Production—'000 tonnes' Yield-kg/ha)

State	1995-96			1996-97			1997-98		
	Area	Production	Yield	Area	Production	Yield	Area	Production	Yield
(1)	(2)	(3)	(4)	(5)	(6)	(7)	(8)	(9)	(10)
Andhra Pradesh	203.7	363.5	1784	262.3	562.0	2143	175.1	303.8	1735
Arunachal Pradesh	1.1	1.4	1273	1.1	1.5	1091	1.3	1.6	1230
Assam	14.1	9.4	667	14.5	9.9	683	14.3	9.5	664
Bihar	7.0	5.3	757	6.1	4.5	738	6.1	5.2	852
Gujarat	18.0	21.0	1167	18.6	18.8	1011	18.5	21.3	11.51
Haryana	3.2	2.5	781	3.3	2.6	788	2.7	2.2	815
Himachal Pradesh	0.9	0.2	222	1.0	0.3	300	1.1	0.3	273
Jammu and Kashmir	0.7	0.5	714	0.6	0.4	667	0.6	0.4	667
Karnataka	188.5	108.1	573	200.8	161.2	803	163.6	130.8	799
Kerala	0.5	0.4	800	0.5	0.5	1000	0.5	0.5	1000
Madhya Pradesh	42.1	16.4	389	48.5	17.4	359	46.6	21.3	458
Maharashtra	102.9	58.5	569	108.2	59.6	551	108.5	60.8	560
Manipur	6.6	4.0	606	7.2	4.3	597	7.2	4.3	597

TABLE-I: Continued

(Area—'000 ha, Production—'000 tonnes' Yield-kg/ha)

State	1995-96			1996-97			1997-98		
	Area	Production	Yield	Area	Production	Yield	Area	Production	Yield
(1)	(2)	(3)	(4)	(5)	(6)	(7)	(8)	(9)	(10)
Meghalaya	1.7	1.1	647	1.8	1.1	611	1.8	1.1	611
Mizoram	2.1	2.0	952	2.8	3.3	1179	2.8	3.3	1179
Nagaland	0.4	3.3	8250	0.4	3.8	9500	0.4	2.7	6750
Orissa	99.5	78.3	787	49.0	40.7	831	92.0	75.0	815
Punjab	2.9	14.4	1517	4.7	7.4	1574	4.7	8.0	1702
Rajasthan	38.5	37.2	966	53.2	59.7	1122	40.5	68.7	1696
Tamil Nadu	69.1	27.7	401	76.7	38.2	498	57.7	28.1	487
Tripura	1.5	0.8	533	1.6	0.9	563	1.9	1.0	526
Uttar Pradesh	20.1	13.5	672	22.4	18.6	830	19.6	16.4	837
West Bengal	58.6	49.8	850	58.9	49.7	844	64.0	55.5	867
Pondichery	Neg	0.1	—	—	—	—	Neg	Neg	Neg
Total	883.7	809.4	916	944.2	1066.4	1129	831.5	821.8	988

Source: D of E and S, New Delhi.

TABLE-II: Area and Production of Chillies

Year	Area (in ha)	Production (in tonnes)
1970-71	783400	520400
1975-76	739800	526100
1980-81	834800	509100
1985-86	904100	877400
1989-90	892600	783300
1992-93	962100	862100
1993-94	930000	800100
1994-95	829100	794700
1995-96	883700	809400
1996-97	944200	1066400
1997-98	831500	821800

Source: D of E and S, New Delhi.

Marketing

The crop becomes ready for harvesting in about 3-4 months after planting. After picking the fruits, they are dried in the sun for 4-5 days, and are graded for size and colour before making. Unripe chillies are sometimes boiled and dried for domestic consumption. Commercially, there are various grades, such as the first sort, the second sort, mixture etc. Grades such as special, medium and fair are also adopted. Good fruit length, shining red colour, high pungency and strong attachment of the calyx are the important factors which the merchants consider for fetching a high price.

The major chilli marketing centres in India are Nasik, Ahmed Nagar, Sholapur, Aurangabad, Nanded, Amaravathi, Lasalgaon in Maharashtra, Guntur, Warrangal, Hyderabad, Cuddapah, Vijayawada, Rajamundri and Nellore in Andhra Pradesh, Dharwad, Mysore, Hassan, Bangalore, Bellary, Ranibennur, Hubli and Byadagi in Karnataka, Pollachi, Ramnad, Madurai, Trichi, Theni, Dindigal, Virudu Nagar and Sathu in Tamil Nadu.

The marketing channels for chillies are of nine types. They are:

(1) Producer farmer — Consumer.

(2) Producer farmer — Retailer — Consumer.

(3) Producer farmer — Wholesaler — Consumer.

(4) Producer farmer — Wholesale — Retailer — Consumer.

(5) Producer farmer — Commission agent — Wholesaler — Consumer.

(6) Producer farmer — Commission agent — Wholesaler — Retailer — Consumer.

(7) Producer farmer — Commission agent to wholesaler — Processing factory — Spice Trader — Selling Agents both internal and external — Consumer.

(8) Producer farmer — Commission agent — Consumer enterprises.

(9) Producer farmer — processor — commission agent outside the country — Retailer — Consumer.

The marketing of chilli remains a major constraint since the commission agents still take a major share of consumer's price. In most of the states where chillies are cultivated, the arrival of chillies are more in an open markets as compared to regulated markets due to many reasons. The main problem faced by the farmers in the regulated market is the time taken for disposing their produce. In these markets, farmers are forced to stay back for more than a day to get remunerative price, because of the nature of bidding in auction, whereas in the open markets, the farmers can have the choice of purchasing outlets and the price is normally fixed by negotiations to a large extent.

The transactions that take place in the regulated market is through the secret tender system. The commodity is arranged in lots, each having separate lot number. The lots are examined by the traders and the price is quoted in the bid slips against the lot numbers, in the tender forms. Then, these tender forms are examined and compared by the market committee authorities and the highest bidder of the particular lot gets it declared. The successful bidder after paying the price to the seller is permitted to lift the produce. In these markets, the traders normally pay through drafts, bills, cheques etc. whereas cash transactions are more or less absent.

The transaction of chillies outside the regulated market i.e., in open market is of speculative nature. The commodity is kept in heaps according to local grades. While executing the deal between the farmer and trader, the commission agent plays a crucial role. The commission agent and the purchaser touch each others fingers by covering their hands with a piece of cloth and the price of the produce is settled. In some of the regulated markets, storage facilities are also available exclusively for the farmers in states like Tamil Nadu, Andhra Pradesh and Karnataka. However, most of the farmers are attracted to open markets because they could get comparative advantage of price and also the choice of channel.

Price Trend

The price of chillies in the markets are decided by the supply of it and also the demand pattern. Table-III gives data for annual average wholesale prices of chillies in important markets in India, which shows that the fluctuations are common in the price in different markets over the period of time.

Export

India exports chillies to the USA, the UK, the Saudi Arabia, Singapore, Sri Lanka and other Asian countries. As far as the import of Indian chillies are concerned, USA stands first in the list followed by Sri Lanka, UAE, Singapore, Malaysia etc. during the period 1997-98. The data for country-wise export of chillies from 1992-93 to 1997-98 is given in Table-IV, which shows that over the years the export of chillies towards the USA market is an outstanding one and the countries like Sri Lanka and Singapore are also importing a large volume of Indian chillies. Again, the demand for Indian chillies in the UK and the Malaysian markets are also increasing since 1992-93. As far as the trend for the export of chillies from India are concerned it was 8,364 tonnes valued at rupees 17,583 thousands in 1960-61 and it came down to 2073 tonnes in 1970-71 worth of rupees 10,867 thousands, which went up in 1975-76 and 1980-81 as can be seen from Table-V, where data is given for the export of chillies from India over the period of time. The minimum volume of 1,241 tonnes of chillies export was noted in 1985-86, however it went forward sharply in 1990-91 to 23,178 tonnes valued at Rs. 2,79,802 thousands, which once again went in the negative direction in 1992-93. The maximum quantum of

TABLE-III: Annual Average Wholesale Prices of Chillies in Important Markets of India

(Rupees/Qtl)

Year	Andhra Pradesh			Tamil Nadu		Karnataka		Maharashtra	Delhi	West Bengal
	Guntur (G.I.)	Warrangal	Hyderabad (Fine)	Tuticorin (SI)	Madras City (Ramnad)	Byadgi (Keddi)	Bangalore	Bombay (Byadgi)	Delhi (Guntur)	Calcutta (Dry)
1993-94	1246	1337	1160	1646	1453	1929	1529	4503	2555	—
1994-95	2544	2226	2026	3141	1973	1714	2542	3284	2782	3499
1995-96	3500	3456	3418	4247	3341	3005	4025	4159	4607	5363
1996-97	2955	2895	2931	3544	3619	3815	4505	6000	4159	4635
1997-98	2119	1389	1949	2697	2871	3097	3125	6178	3682	3485
1998-99	3485	3803	3726	4752	4290	N.A.	4330	—	4910	5669

Source: D of M'g. AP, Hyderabad / ASI Tuticorin / COS Madras.
APMC Byadgi, RMC. Bangalore, AD of AM Calcutta.
M'g officer Delhi, DD of M S and R Maharashtra, Bombay.

TABLE-IV: Country-wise Export of Chillies from 1992-93 to 1997-98

(Quantity—tonnes, Value—'000 Rs.)

Country	1992-93		1993-94		1994-95		1995-96		1996-97		1997-98	
	Qty.	Value	Qty.	Value	Qty.	Value	Qty.	Value	Qty.	Value	Qty.	Value
UAE	603	14909	2651	61232	3388	96430	10731	339681	3852	137650	2498	71510
Sri Lanka	1733	78383	6508	124114	6404	147751	9050	309682	5154	184319	2501	70376
Nepal	46	1665	19	105	2	82	9	193	69	265	637	7504
USA	6894	300600	5996	169405	4058	128660	7666	335915	9205	420655	3703	170545
Singapore	1215	24677	930	16226	740	20151	3887	142681	7279	280876	2344	53373
Malaysia	23	683	431	7516	209	6250	2153	80532	5892	244340	1804	41907
Bangladesh	77	2594	—	—	383	8765	10863	334869	154	6822	—	—
USSR	118	3006	207	11680	35	1135	169	8438	241	12848	695	4084
Saudi Arabia	270	5644	591	12609	253	6406	576	19463	913	29865	490	14536
Total (including others)	15144	606889	30776	721357	20096	571165	56165	1954620	50051	2014517	42172	1388463

Source: DG of CI and S, Calcutta.

chillies export from India in recent years was observed during the period 1995-96 that of 56,165 tonnes valued at Rs. 19,54,620 thousands but in the subsequent years it came downwards.

TABLE-V: Export of Chillies from India

Year	Quantity (Tonnes)	Value ('000 Rs.)
1960-61	8364	17583
1965-66	9532	24871
1970-71	2073	10867
1975-76	3532	31806
1980-81	7682	55559
1985-86	1241	20203
1990-91	23178	279802
1992-93	15144	606889
1993-94	30776	721357
1994-95	20096	571165
1995-96	56165	1954620
1996-97	50051	2014517
1997-98	42172	1388463
1998-99	55750	2101300
1999-2000*	50000	1750000

*Target

Source: DC, CI and S, Calcutta.

India is also an important exporter of oleoresin of chillies to the international market. The main importers of this are Australia, France, Germany, Japan, the UK and the USA. During the period 1997-98, the USA imported the highest volume of chilli oleoresin from India fcllowed by Japan and German FR as can be observed from Table-VI. The demand for oleoresin of chillies in the German market was the highest one during the period 1992-93 and 1993-94 but from then onwards it came downwards sharply and the same is the case with the UK market too. As far as the total volume of oleoresin of chillies export since 1992-93 is concerned, the highest volume of 1,05,002 kgs was noted in 1997-98 while in 1996-97 it was only 11,703 kgs, which was the lowest one since 1992-93 upto 1997-98.

TABLE-VI: Country-wise Export of Oleoresin of Chillies from India

(Quantity in kg., Value in '000 Rs.)

Country	1992-93		1993-94		1994-95		1995-96		1996-97		1997-98	
	Qty.	Value	Qty.	Value	Qty.	Value	Qty.	Value	Qty.	Value	Qty.	Value
Australia	3430	2144	7911	4637	3000	1386	—	—	660	503	600	311
France	121	97	1970	1733	1540	1339	1200	485	720	975	300	420
German FR	29342	31212	18816	22537	702	704	490	554	1345	1133	1085	1374
Japan	2112	3391	22500	30585	18670	24775	—	—	510	833	15461	18566
Switzerland	—	—	—	—	10	6	—	—	—	—	—	—
UK	5560	2216	6654	5998	8287	6954	1000	399	1540	1002	218	223
USA	12983	15178	5300	4908	1131	478	100	176	1393	1155	60231	49205
Total (including others)	69756	77569	103803	10889	51633	51532	14282	12677	11703	16594	105002	105327

Source: DG of CI and S Calcutta.

As a whole, there appears the ups and downs in India's export of chillies and oleoresin of chillies over the years, which is not a positive sign as far as the development of the sector in particular and that of the nation in general is concerned. This type of fluctuation in the exports appear to be there because of the problems of this sector. They are:

(1) Fluctuation, stagnation or low yield in chillies is the main factor for a change in our exports. This has been due to the cultivation of disease-susceptible varieties, especially the virus susceptible chillies. A large number of production constraints limit chilli production in the country. The constraints are—Incidence of leaf curl and bacterial wilt, lack of varieties high in Capsaicin, Oleoresin and capsanthin and absence of hybrid technology. The yield in India is much lower as compared to Indonesia, Brazil and Malaysia.

(2) Among the production constraints, incidence of leaf curl and bacterial wilt are major problems. A large number of complex diseases like pepper virus complex, leaf curl complex, Rapid decline, Mosaic complex and Moria disease limit chilli production in India. Various studies on chillies shows that leaf curl leads to 100 per cent loss and fruit rot by Alternaria leads to 50 per cent of loss.

(3) Marketing of chilli is a major constraint. In the major chilli marketing centres in the state of Maharashtra, Andhra Pradesh, Karnataka and Tamil Nadu shows that the commission agents still take a major share of consumer's price.

(4) Cost of cultivation in chilli is going up in recent years due to high cost of labour and plant protection chemicals.

(5) The cost of transportation is ever increasing.

(6) Non-availability of proper curing methods.

(7) The main reason for the declining exports is high prices as compared to Pakistan and China who are the main competitors for India.

(8) Our chilli exports suffer because of adulteration and poor packing.

Prospects

India has immense potentiality to grow chillies. The world demand is also going up for chillies, which is more than 40,000 tonnes. The volume of exports of chillies, chilly powder and oleoresin of chillies is too small in relation to the world production and demand. As India has been traditionally exporting these, there is a vast scope for increasing this. In this regard there is an urgent need for a long-term strategy, which should contain the following important aspects:

(1) There is the need to identify varieties resistant to various diseases, especially to leaf curl complex.

(2) There is the need to come up with a set of agronomic practices to increase the productivity. In this regard control measures against dieback, fruit rot, powdery mildew and bacterial leaf spot has to be taken. Again control measures have to be worked-out against major pests of chilli like "Podborer" and "Pest complex".

(3) Need to identify high yielding and processing oriented varieties which should contain more of oleoresin, colour and capsaicin. This can meet the requirements of the chilli growers, traders, consumers and industrialists.

(4) So as to overcome the problem of marketing and that of the commission agents, the only solution is co-operative marketing which is yet to take a start in chilli arena.

(5) For improving exports improved flavour, pungency, vitamins and quality of oleoresin has to be built into the chilli varieties.

(6) So as to minimise waste as and when production is more, curing should be done by the "Netsack method" which is practised in Hungary.

(7) The world trade in Paprika Oleoresin has shown an upward trend in the recent years, so there is an imperative need to develop high yielding Paprika like chillies, mildly pungent and having high colour value as there is a great demand for such varieties in the international market.

Above all, agrotechniques seed-production techniques and development of varieties resistant to biotic and abiotic stress and ideal processing methods in this crop have to be standardised. Then only, India can hope to gain the major external markets like West Asia, North Africa, Russia, EECs, (East European Countries) USA, Canada, Japan and Australia for chilli and its products.

5

GINGER

Ginger is the dried underground stem or rhizome of the Zingiberous, herbaceous plant Z. officinale, which constitutes one of the most important major spices of India. The historical records show that ginger was certainly known to and highly esteemed by the ancient Greeks and Romans who obtained this spice from Arabian traders via the Red Sea. It was introduced into Germany and France in the ninth century and to England in the tenth century, later the plant had been introduced to many tropical and sub-tropical countries.

Ginger is now cultivated in several parts of the world, the important countries producing this are Thailand, Japan, Jamaica, Indonesia, Nigeria, China, Bangladesh, Fiji, Korea and Australia. India is the largest producer of ginger in the world. In 1980-81 India's share in the world's production of ginger was 65 per cent and at present it is about 35 per cent. In recent years countries like Thailand, Japan, Bangladesh, South Korea and Indonesia are emerging as the leading ginger producing countries in the world.

Area, Production and Yield of Ginger in India

In India the cultivation of ginger is done in almost all the states. The total area under ginger at present is around 68 thousand hectares and the production is about 233.66 thousand tonnes, while the yield per hectare is 3477 kg. Among the states where ginger is grown, Kerala alone possesses about 25 per cent of the total area under it and its production is also accounting for the same percentage. This is being followed by West Bengal, Arunachal Pradesh, Meghalaya, Orissa, Sikkim and others. In terms of yield per hectare—Tamil Nadu, Nagaland, Meghalaya, U.P., Arunachal Pradesh, Kerala etc., are above the all India level. The figures for the above aspects are given in Table-I.

The overall trend in area, production is shown in Table-II. Accordingly the total area in 1970-71 under ginger was 21590 hectares and the production was 29290 tonnes which was increased to in the subsequent years and has reached to 53520 hectares and 138020 tonnes in 1985-86. From there onwards, the area under ginger and its production trend clearly shows the positive sign of it's growth in India.

Ginger crop is grown from almost the sea-level up to an altitude of 1,500 metres. It comes up well on a variety of soils, provided sufficiently well-distributed rainfall or irrigation and adequate drainage facilities are available. In irrigated lands, it is rotated with betel-vine, plantain, turmeric, onion, garlic, vegetables etc. On coconut, coffee and orange plantations on the west coast, ginger is grown as an inter-crop. In Himachal Pradesh, tomato and chilli are grown as an inter-crop with ginger.

Varieties

There are several varieties of ginger grown in different parts of India. Important among them are Maran, Nadia, Wynad, Rio-de-Janeiro, China, Poona, Karakal, Ernad, Thingpui etc. Of these Maran, Nadia and Karakal have higher dry ginger recover while for vegetable ginger Rio-de-Janeiro, China, Wynad etc., are preferred. All these varieties differ in shape, size, rhizomes, yields, moisture content, flavour, quality etc. For export purpose Cochin and Calicut gingers are graded according to the number of fingers contained in the rhizome. Recently introduced improved varieties are Suprabha, Suruchi, Surabhi, Himgiri, IISP Varada, through these varieties the expected yield is about 4000-6000 kg/ha. In the domestic as well as the international market the demand and price of this is mainly determined by its variety, quality and contents.

TABLE-I: State-wise Area, Production and Yield of Ginger in India

(Area '000 Ha, Production in '000 Tonnes, Yield in kg/ha)

State	1995-96			1996-97			1997-98		
	Area	Production	Yield	Area	Production	Yield	Area	Production	Yield
Andhra Pradesh	2.39	8.87	3711	2.41	8.96	3718	1.91	6.77	3545
Arunachal Pradesh	2.79	19.45	7125	3.33	24.34	7201	4.19	32.08	7656
Bihar	0.70	0.92	1314	0.79	1.13	1430	0.80	1.12	1400
Gujarat	0.49	1.96	4000	0.56	2.24	4000	0.65	2.60	4000
Haryana	0.03	0.01	333	0.03	0.01	333	0.04	0.01	250
Himachal Pradesh	1.38	0.66	478	1.43	0.69	482	1.43	0.69	482
Karnataka	3.51	4.65	1325	4.41	5.81	1317	3.57	4.73	1324
Kerala	14.03	50.61	3607	12.81	48.47	3783	13.52	51.72	3825
Madhya Pradesh	3.01	4.05	1346	3.70	5.36	1449	3.77	5.03	1334
Maharashtra	1.06	1.01	953	1.06	1.01	952	1.24	1.19	959
Manipur	0.74	1.22	1649	0.74	1.22	1649	0.74	1.22	1649
Meghalaya	7.29	42.96	5893	7.31	46.18	6317	7.36	45.26	6149
Mizoram	1.54	14.58	9468	4.37	21.79	4986	2.63	20.43	7768

Continued...

TABLE-I: Continued

State	1995-96			1996-97			1997-98		
	Area	Production	Yield	Area	Production	Yield	Area	Production	Yield
Nagaland	0.43	6.38	1484	0.47	7.20	15319	0.50	4.00	8000
Orissa	11.67	17.80	1525	9.83	10.56	1074	7.14	7.68	1075
Rajasthan	0.30	0.81	2700	0.28	1.09	3893	0.32	0.65	2031
Sikkim	4.50	3.84	853	4.60	4.00	869	4.94	4.23	856
Tamil Nadu	0.70	15.90	22714	0.60	14.50	24166	0.61	14.95	24508
Tripura	0.99	1.82	1838	1.01	1.85	1832	1.01	1.85	1832
Uttar Pradesh	0.90	3.98	4422	1.32	7.00	5303	1.30	7.44	5723
West Bengal	8.13	16.69	2053	8.85	17.94	2027	9.14	18.85	2062
Andaman & Nicobar	0.37	1.13	3054	0.38	1.16	3053	0.39	1.16	2974
All India	65.98	209.88	3279	70.29	232.51	3308	67.20	233.66	3477

Source: DG S and S, New Delhi.

TABLE-II: Area and Production of Ginger in India

Year	Area (in Hectares)	Production (in Tonnes)
1970-71	21590	29290
1975-76	27200	45150
1980-81	40450	82440
1985-86	53520	138020
1989-90	53020	152890
1992-93	59870	201630
1993-94	60580	186200
1994-95	61090	197650
1995-96	65980	209880
1996-97	70290	232510
1997-98	67200	233660

Source: D of E and S, New Delhi.

Products of Ginger

The quality of products of ginger depends on the variety used and cultivation practices.

Raw/Dry ginger

Himachal, Maranthady, Kuruppampady etc., are good varieties to prepare dry ginger. Rio-de-Janeiro, China, Wynad, Varada etc., are good varieties for raw ginger. The major Indian trade types are Cochin and Calicut ginger. Cochin ginger is more superior in quality. Appearance, contents of volatile oil and fibre, pungence level and a subjective assessment of aroma and flavour are important in the quality evaluation of dried ginger. This depends on the cultivator, stage of maturity at harvest etc.

Ginger oil

It varies from 0.5 to 3.0 per cent oil possesses only the aroma and not the flavour of the spice.

Ginger Oleoresin

It is a blend of oil and resinoides. It is extracted from ginger powder using organic solvents like acetone, ethylene dichloride etc.

Fresh ginger products

Ginger preserve of Murabba, ginger candy, soft drinks like ginger cocktail, ginger pickles, salted ginger, salted in vinegar or vinegar mixed with lime, green chillies etc.

Uses

Ginger has a distinctive, spicy, penetrating flavour, pungent and slightly biting due to antiseptic or pungent compounds present in it, which make it indispensable in the manufacture of a number of food products like ginger bread, confectionery, ginger ale, curry powders, table sauces, in pickling and in the manufacture of certain soft drinks etc. Ginger is also used for the manufacture of ginger oil, oleoresin, essences, tincture etc. A number of alcoholic beverages are prepared from ginger in foreign countries such as ginger brandy, ginger wine, beer etc.

In the Ayurvedic medicinal system, ginger is considered to be carminative, stimulant and given in dyspepsia and flatulent colic. It is also prescribed as an adjunct to many tonic and stimulating remedies.

The oil of ginger finds use in perfumery, where it imparts a unique individual note to compositions of the oriental type.

Marketing

The ginger harvested is sold as green ginger to the middlemen or to the local merchants and sometimes through the sub-dalals and dalals, agents of commission, agents of secondary market or wholesale terminal market. There is no organized market for ginger in most of the growing states. As far as dry ginger is concerned, there are a number of marketing centres and agencies for the distribution of it in the country and for the exports also. The marketing agencies are managed mainly by the private traders. A few cooperative marketing societies have been handling this product in small quantities. Absence of effective competitive market for ginger, has made the way for entry of private traders and they are able to make profits through manipulation of price. The main marketing channels for the marketing of ginger from the point of view of the producers are:

(1) Producer — Village merchant — Commission agent — Exporter — Consumer.

(2) Producer — Village merchant — Commission agent — Wholesaler of assembling markets — Wholesaler of consuming market — Retailer — Consumer.

Price Behaviour

For the development of the ginger industry, it calls for a reduction in the price spread in ginger marketing. Several studies were conducted time and again so as to notedown the pattern of price spread in ginger marketing, almost all of them shows that the producer receives about 45-64 per cent of the price paid by the ultimate consumer in the consignment exported and that sold in the internal markets. The share of the marketing costs is around 25 per cent in the export trade and 27 per cent in the internal market while that of marketing margins between 12-13 per cent in the former and 29 per cent in the latter. Data for average annual wholesale prices of ginger in the important markets in India are given in Table-III which reflects that there are ups and downs in the level of prices over the years in different markets.

Exports

The demand for ginger and its products are ever increasing in the international market. India exports a large quantum of ginger to the countries like Pakistan, Saudi Arabia, USA, UK, UAE, Bangladesh, etc. The high income oil exporting countries have been the major importers of ginger from India over the years. Nearly 60 per cent of our export of ginger goes to Pakistan followed by Bangladesh, USA and UAE. In the earlier eighties, Saudi Arabia alone use to import nearly 40 per cent of Indian ginger. In the nineties our neighbouring countries like Pakistan and Bangladesh were importing the major portion of our ginger, as can be seen from Table-IV while in the seventies and eighties Saudi Arabia, Kuwait and UAE use to import nearly 52 per cent of our ginger.

As far as our ginger export trend over the years is concerned in 1960-61 it was 5577 tonnes worth Rupees 8372 thousands, which

TABLE-III: Average Annual Wholesale Prices of Ginger in Important Markets of India

(Rs./Qtl)

Year	Tellechery (Dry)	Calicut (Dry blended)	Cochin (Unbleached)	Bombay (Bleached)	Madras (White)	Guwahati (Wet)	Calcutta (Wet)
1993-94	2538	2673	2304	3107	3371	999	—
1994-95	4318	4622	4846	5073	5375	1409	1523
1995-96	5924	5942	6199	6865	7208	1488	1531
1996-97	4273	4511	4845	7951	6352	1007	1013
1997-98	4374	4820	5200	7957	6476	845	1565
1998-99	4956	5276	5277	—	8150	1196	1799

Source: Price Reporter/ Inspector, Market Intelligence, Cochin Regional Officer, Marketing Intelligence, Kozhikode, Delhi Administration Delhi, Commissioner of Statistics Madras, Deptartment of Agriculture Marketing Guwahati, Assam ADD of Agriculture Marketing, West Bengal Calcutta Assistant Director of Marketing (Prices) Bombay.

TABLE-IV : Country-wise Export of Ginger from India

(Qty in Tonnes, Value '000 Rs.)

Country	1995-96		1996-97		1997-98	
	Qty.	Value	Qty.	Value	Qty.	Value
S. Arabia	1030	73113	872	47990	1067	60052
Yeman FD Republic	681	44800	834	44610	873	44618
USA	354	19657	1261	54516	1609	52584
USSR	—	—	—	—	10	621
Morocco	379	23439	460	26381	615	36437
Kuwait	23	1555	38	2056	116	7891
Pakistan	8072	98530	19990	256006	17032	243865
UK	214	11739	270	13836	259	17043
UAE	263	13664	569	26396	1438	55500
Singapore	—	—	44	2402	210	10923
Netherlands	161	12101	402	238	297	17661
Bangladesh	6632	58209	3688	29284	2394	13053
Total (including others)	18483	389213	29737	592441	28310	726272

Source: DG of CI and S, Calcutta.

come down to 3156 tonnes in 1970-71, from then onwards it has been increasing with slight variations as can be noted from Table-V. In the year 1996-97 we exported 29737 tonnes of ginger, which was considered as the highest one in recent years valued at Rupees 592441 thousands.

TABLE-V: Export of Ginger from India

Year	Quantity (M.T.)	Value (Rs. '000)
1960-61	5577	8372
1965-66	3987	13249
1970-71	3156	26094
1975-76	4786	41049
1980-81	6811	36797
1985-86	6816	108935
1990-91	5487	109339
1992-93	9825	168737
1993-94	18442	247810
1994-95	12022	167303
1995-96	18483	389213
1996-97	29737	592441
1997-98	28310	726272
1999-2000*	14000	700000

*Target
Source: DG of CI and S, Calcutta.

India export ginger oleoresin and oil on larger volume. Ginger oleoresin is obtained by extraction of powdered dried ginger with suitable solvents like alcohol, acetone or any other efficient solvent. Unlike volatile oil, it contains both the volatile oil and the non-volatile pungent principles for which ginger is so highly esteemed. Concentration of the acetone extract under vacuum and on complete removal of even traces of the solvent used, yields the so called oleoresin of ginger. Ginger oleoresin is manufactured on a commercial scale in India and abroad and is in great demand by the various food industries. While ginger oil is obtained by steam distillation of dry ginger powder. India export these two products to countries like USA, UK, Canada, German FRP, Netherlands,

Japan, Australia etc. As far as ginger oleoresins are concerned, USA is the main market for it followed by UK, German FRP and Canada. In 1997-98 USA imported 16483 kgs of this valued ginger at Rupees 17699 thousands whereas we exported 13577 kgs of it to UK during the same period. From a mere 50 kgs of export of oleoresins in 1973-74 it went up to 67978 kgs in 1997-98 valued at Rupees 73284 thousands and trend over the years as shown in Table-VI reveals that it has been moving in the positive directions. The same is the case with the export of ginger oil too. With regard to ginger oil export USA is the major market for us followed by German FRP, Japan and Netherlands. USA imported 4128 kgs of this in 1997-98 worth of Rupees 10275 thousands where as German FRP's import was 2483 kgs during the same period.

In terms of total export of ginger oil in 1973-74. It was just 114 kgs. worth of 45 thousands rupees but from then onwards it is moving in the right direction as can be observed through Table-VI.

TABLE-VI: Export of Ginger Oleoresins and Oil from India

(Quantity in kgs. Value in Rs. '000)

Year	Ginger Oleoresins		Ginger Oil	
	Qty.	*Value*	*Qty.*	*Value*
1973-74	50	5	114	45
1983-84	9193	3305	4732	3838
1993-94	43243	34688	3867	4371
1994-95	53529	49552	5825	8089
1995-96	56732	54744	5258	14200
1996-97	45307	45061	4115	10956
1997-98	67978	73284	10381	34635

Source: DG of CI, Calcutta.

The above aspects clearly indicate that there is a bright future for this sector since the demand in the world market for ginger and ginger products are ever increasing. But it is really heartening to note that India has not succeeded in obtaining the position which was theirs, prior to 1985. If India wants to regain that position, there

is the need to overcome the prevailing problems of this sector. They
are:

1. The production and productivity of ginger in India is much
   lower as compared to Thailand, China, etc., this is because
   of the non-availability of qualitative high yielding varieties
   of seeds. The prevailing seeds are not based upon the
   different agroclimatic conditions and are used by the
   farmers without knowing about the productivity.

2. Problem of pests and diseases limit the production, pests
   like shoot borer, leaf roller, scale insecticides etc., and
   diseases like soft rot, leaf spot are responsible for the
   variation in production and productivity.

3. Drying of ginger in many states appears to be non-practical
   due to economic and climatic conditions especially in the
   north-eastern region as the harvesting period of ginger
   synchronises with mid-winter period and thus, sun drying,
   which is the most economical method followed in Kerala,
   is not possible. Because of this the ginger becomes
   unacceptable for exports.

4. The prevailing marketing channels actually deceive the
   farmers. Absence of organized marketing system is a major
   hurdle of the sector.

5. Absence of grading in most of the growing states except in
   Himachal Pradesh limits the scope for an improvement in
   the quality of ginger for exports.

6. Intense competition in the external market from countries
   like China, Taiwan, Thailand, South Korea etc.

Prospects

The above problems has to be solved without much delay since
ginger is consumed both internally and externally in an increasing
manner. So as to solve these, a planned long term strategy is the
need of the day and it should contain the following aspects.

1. There is the need to improve the production and
   productivity with the help of high yielding short-term
   varieties. The research wings should identify these type of

varieties which could be ideal for different agroclimatic conditions. Along with this, proper education and training has to be given to the farmers with regard to various aspects of cultivation, fertilization, harvesting and post-harvesting activities. Priorities should be given for the evolution of short duration ginger varieties. Besides, the breeders should aim to evolve varieties with high yield, low fibre content, high oil and oleoresin in ginger.

2. Timely information should be given to the farmers along with required pest and disease control devices. Efforts are needed to take the available technical know-how to the common farmers which alone will help to reduce the yield gap and enhance production without any time lag.

3. More research and developmental activities should be conducted so as to provide low cost mechanical devices and technology for drying and grading. There is also the need to conduct marketing and processing research which will be helpful in the preparation of policy matters. Research on by z-product utilization and product diversification should be undertaken.

4. Need to improve the marketing facilities with proper storage and godown facilities. Proper post-harvest storage techniques may be evolved for safeguarding the final crop from spoilage. So as to boost our exports of ginger, there is the need to reduce the cost of production which ultimately reduce the price of our ginger and thereby competitiveness improves. Along with this, there is also the need to reduce the fibre content through high-yielding, disease-resistant varieties which are richer in volatile oil and oleoresin content. There is also the need to diversify the products and markets to improve the exports.

5. Agricultural Produce Market Act should be introduced and enforced for ginger in states like Kerala. The farmers should be educated for selling their produce in the regulated markets.

6. It is necessary to strengthen the existing cooperative marketing societies and establish new ones where they do not exist to provide adequate and timely credit to

producers to free from the hold of moneylenders and commission agents.

7. Need to compile data on the production and related aspects.

The domestic consumption of ginger is ever increasing over these years in India and the available quantum for export is minimum, which is below 25 per cent, hence efforts are needed to promote the production and productivity in India. Again the producers as well as traders should concentrate on impressing. Improving the quality of our ginger then only we can hope for better days.

6

TURMERIC

Turmeric is the dried rhizome of Curcuma longa L, a herbaceous perennial belonging to the family Zingiberaceae and a native of India or China. It is an important condiment and a useful dye, with varied uses in drug and cosmetic industries. It is used medicinally for external application and taken internally as a stimulant. It is cultivated extensively in India, Sri Lanka, Indonesia, China, Peru, Jamaica and other tropical and subtropical countries. India is by far the largest producer of turmeric in the world.

Area, Production and Yield of Turmeric in India

In India, turmeric is cultivated in Andhra Pradeeh, Orissa, West Bengal, Assam, Maharashtra, Tamil Nadu, Karnataka and Kerala on a large scale while in the other states in a small scale. At present the total area under turmeric is around 125 thousand hectares and the production is about 488 thousand tonnes and the yield is 3912 kg/ha.

Andhra Pradesh ranks first both in area and production of turmeric in India followed by Tamil Nadu, Orissa, West Bengal, Assam etc., as can be seen from Table-I which gives data on area, production and yield of turmeric in different states for the period 1994-95 to 1997-98. The figures shown in the Table-I give an idea that in terms of area, production as well as yield, has been fluctuating over the years.

As far as the total area under this crop was concerned in 1970-71 it was 80500 hectares and the production was 150600

TABLE-I: State-wise Area, Production and Yield of Turmeric

(Area—'000 ha, Production—'000 tonnes, Yield—kg/ha.)

State	1996-97			1997-98		
	Area	Production	Yield	Area	Production	Yield
Andhra Pradesh	52.0	292.0	5615	48.0	273.0	5688
Arunachal Pradesh	0.5	1.2	2400	0.3	1.0	3333
Assam	10.0	6.9	690	10.1	7.0	693
Bihar	3.4	3.3	971	3.1	3.6	1161
Gujarat	0.4	11.0	2750	0.4	7.4	18500
Karnataka	4.8	26.2	5458	4.5	25.0	5596
Kerala	3.8	9.1	2395	3.6	8.4	2333
Madhya Pradesh	0.7	0.7	1000	0.7	0.6	857
Maharashtra	7.2	9.2	1278	7.3	9.1	1247
Meghalaya	1.4	6.4	4571	1.4	6.4	4571

Source: D of E and S, New Delhi.

TABLE-I: (Contd.)

(Area—'000 ha, Production—'000 tonnes, Yield—kg/ha.)

State	1996-97			1997-98		
	Area	Production	Yield	Area	Production	Yield
Mizoram	0.4	3.6	9000	0.4	3.6	9000
Orissa	19.7	38.5	1954	14.3	28.0	1958
Rajasthan	0.2	0.8	4000	0.2	1.4	7000
Sikkim	0.3	1.0	3333	0.4	1.2	3000
Tamil Nadu	16.0	92.0	5758	15.2	83.7	5507
Tripura	1.4	2.8	2000	1.4	2.8	2000
Utter Pradesh	0.4	1.7	4250	0.8	1.6	2000
West Bengal	12.6	22.5	1786	12.5	23.6	1888
All India	135.2	528.9	3912	124.6	487.4	3912

Source: D of E and S, New Delhi.

tonnes and this came down upto 1997-98 and in 1978-79 in terms
of area it went up to 89700 hectares with a production of 190400
tonnes. From then onwards both in terms of area and production
it shows certain positive response eventhough there appeared
slight variations in these as can be seen from Table-II.

TABLE-II: Area and Production of Tumeric in India

Year	Area (in Ha)	Production (in Tonnes)
1970-71	80500	150600
1975-76	71800	135200
1980-81	101500	216900
1985-86	109300	367100
1989-90	123900	390000
1992-93	130200	407700
1993-94	148400	707400
1994-95	149400	622000
1995-96	139300	462900
1996-97	135200	528900
1997-98	124600	487400

Source: D of E and S, New Delhi.

Turmeric cultivation requires a warm and humid climate. In
heavy rainfall tracts of the west coast, it is grown as a rainfed crop
and in the other states it is cultivated under irrigation. In wet lands
turmeric is rotated with paddy, sugarcane, banana etc., once in 3
or 4 years. In garden lands it is grown in rotation with sugarcane,
chilli, onion, garlic, vegetables, wheat, ragi etc. In Gujarat, it is
cultivated as a crop subsidiary to ginger and in other areas with
chilli and quick growing vegetables. In West Bengal, it is grown as
an intercrop with mango, litchi, and jackfruit and on the west coast
with coconut and arecanut (Katha).

Varieties

Under the genus Curcuma to which turmeric belongs, the
botanists have so far recognised more than 30 varieties of these,
C.longa is economically the most important accounting for about
96 per cent of the total area under turmeric and the remaining of

the total area are under C. Aromatica, which is mostly grown in small areas in East and West Godavari districts of Andhra Pradesh and Tanjavoor and South Arcot districts of Tamil Nadu. Turmeric produced in different areas is known by various local names. Alleppy turmeric is reported to contain percentage of curcumin and is considered one of the best in the World. The Patna variety is noted for its deep colour, of the two types cultivated in Maharashtra, 'Lokhandi' has bright coloured hard rhizomes and the other has light-coloured soft rhizomes. The popular commercial varieties in Andhra Pradesh are 'Duggirala' of Guntur, 'Kasturi Pasupa' of Godavary Delta etc. The improved varieties available at present are Ccl, Krishna, Suroma, Ranga, Suguna IISR Prabha, IISR prathiba, BSR-1, BSR-2, Megha turmeric I etc., and the expected yield from these improved varieties varies in between 4000-6000 kg/ha. Depending upon the varieties the crop is ready for harvesting in about 7 to 9 months after sensing. The curing quality and the proportion of the cured and dried produce to the green produce depends mainly on the variety. The quality of the final product, including its colour and aroma depends largely on the right amount of curing.

Uses

Turmeric is a unique, colourful and versatile natural plant product combining the properties of a spice, colourant, as a cosmetic and as a drug. However, it is largely consumed as a spice. A larger volume of turmeric is utilised in most of the Asiatic countries as a food adjunct in many vegetables, meat and fish preparations. It is used to flavour and at the same time to colour butter, cheese, pickles and other food stuffs. It is also used to colour liquor, fruit drinks, cakes etc.

Earlier in India, turmeric was largely used for dyeing wool, silk and cotton to impact a yellow shade, in an acid bath. The dye is also employed as colouring material in pharmacy, confectionery and food industries. Considerable qualities of turmeric are converted as 'Kumkum'.

In the Indian system of medicine, turmeric occupies an important place, as an ingredient in the preparation of medicinal oils, ointments etc. It is a stomachic, carminative, tonic, blood purifier, vermicide and an antiseptic. Its antiseptic and healing properties are said to be both a preventive cure for that much-feared malady of adoloscense-pimples. It is also known to discourage unwanted hair on the feminine skin.

Marketing

In the marketing of turmeric internally three type of intermediaries appear before it reaches the consumer from the cultivator, they are village merchants, processors and retailers. Apart from these regulated markets are also involved in the marketing of turmeric especially in Andhra Pradesh. In Tamil Nadu also there is an organized marketing system for turmeric. The absence of an organized marketing system in most of the states adversely affects the growers and thereby the prices fluctuate frequently. The basic facilities like drying, grading, storage etc. are more or less absent in the marketing centres.

Price Behaviour

The arrivals of turmeric in the post-harvest period is normally high because of the lack of holding capacity of the farmers and thereby the prices are generally low. Further, the instability in incomes of the farmers forces them to sell the product at a lower price to the village merchants. The arrival of turmeric are higher during the month of March-June in both bulbs and fingers, and during these months the prices are lower. On the other side from September-February the arrivals are lower and prices are normally higher. As the harvesting of the crop starts from March, market arrivals gradually rise from this month, reaches a peak during April and May, hence the price go downwards. As a whole, the annual average wholesale prices of turmeric in the important markets of India as shown in Table-III, showing that the price of this has been picking up in recent years. In some of the markets the price of turmeric is comparatively more than the others because of high curcumin content of the turmeric traded and the low moisture content.

TABLE-III: Annual Average Wholesale Prices of Turmeric in Important Markets in India

(Rs./Qtl.)

Year	Andhra Pradesh						Tamil Nadu Madras		Maharashtra		Kerala Cochin	Delhi N. Delhi	West Bengal Calcutta Rajmondhi
	Cuddapah		Duggirala		Nizamabad				Sangli Spot	Bombay Rajapuri			
	Finger	Bulb	Finger	Bulb	Finger	Bulb	Finger	Bulb			Finger	Finger	Finger
1993-94	1405	1389	1443	1434	1354	1347	2662	1983	1333	2344	2092	2111	—
1994-95	908	848	864	857	862	833	2031	1313	1040	1927	2189	1822	1057
1996-97	1081	1001	1009	978	891	874	2157	1531	1249	1862	2257	1796	918
1997-98	1899	1806	1818	1616	1679	1614	5912	2715	1922	2913	2788	2766	2085
1998-99	—	—	2358	2263	2380	2149	4583	3768	—	—	—	3830	3813

Source: D of M'g, AP, Hyderabad, Commissioner of Statistics, Madras Assistant Director of M'g (Spices) Bombay

Secretary, Spices and Oilseeds Exchange Ltd., Sangli Price Inspector (MI) Cochin A.D. of Agriculture (M'g) Calcutta.

TABLE-IV: Country-wise Export of Turmeric from India

(Qty. in tonnes, Value in '000 Rs.)

Country	1995-96		1996-97		1997-98	
	Qty.	Value	Qty.	Value	Qty.	Value
USA	2228	68484	2079	76756	2114	97228
UK	1719	34863	1906	53795	1517	54573
Iran	3314	49022	2269	43817	3960	101283
Japan	1761	31322	2046	58001	1956	67515
Singapore	789	15593	430	11370	512	16389
Baharin	230	4239	65	2004	144	5680
UAE	5330	64759	4104	88563	4687	114221
S. Arabia	279	5501	357	10391	292	33196
Canada	164	3701	141	5266	205	8245
Morocco	456	7703	105	2224	406	11695
Total (including others)	27050	462032	23019	584458	26838	791491

Source: DG of CI and S, Calcutta.

Export

Turmeric is one of the most important and ancient spices of India as well as a traditional item of export, it ranked third among the spices exported front India being next to black pepper and cardamom. India exports turmeric to USA, UK, Iran, Japan, UAE, Singapore, Canada, Morocco etc. At present UAE is the major consumer of our turmeric followed by Iran, USA, Japan and others. In 1997-98, the total volume of our turmeric export was 26838 tonnes valued at Rupees 791491 thousands, this was 23019 tonnes valued at Rs. 584458 in 1996-97, as can be seen from Table-IV. As far as the total volume of turmeric export is concerned it has been increasing over the years with slight variations as can be observed from Table-V.

TABLE-V: Export of Turmeric from India

Year	Quantity (M.T.)	Value (Rs. '000)
1960-61	2310	2584
1965-66	10403	13754
1970-71	11109	38347
1975-76	11755	42119
1980-81	14517	78824
1985-86	8562	120944
1990-91	12767	143390
1992-93	19724	488543
1993-94	25436	525600
1994-95	28256	451795
1995-96	27050	462032
1996-97	23019	584458
1997-98	26838	791491
1998-99	36522	1245500
1999-2000*	29500	880000

*Target

Source: DG CI and S, Calcutta.

Turmeric oleoresins has greater demand in USA, UK, Canada, Australia, etc. Indian export of this in 1997-98 was 157510 kgs worth of Rupees 130745 thousands which was 154687 kgs in 1996-97 valued at rupees 86733 thousands. As far as country-wise exports of turmeric oleoresins from 1994-95 to 1997-98 is concerned, the data is given in Table-VI.

From the above aspects on our export sector it is evident that we have been moving in the positive direction. However, when the total volume of turmeric export is compared with the total production, the percentage of it is much lower. This is mainly because of the increasing level of domestic consumption. These factors shows that there are ample opportunities for this crop in the future both in the domestic as well as in the external market, if the existing problems are tackled properly. These problems are:

1. Lower level of production and productivity are the main problems of this sector again year after year the area under this is fluctuating, which really reduces the production.

2. The prevailing system of cultivation is traditional and the farmers are not knowing the modern methods of cultivation.

3. Problem of insect and pests. Turmeric shoot borer and turmeric skipper are the major pests.

4. The prevailing marketing system is exploiting the growers, by way taking a major portion of the consumers rupee.

5. Absence of grading, standardization, non-availability of storage and godown facilities are the major hurdles of this sector.

6. Lack of interest from the side of traders in improving the exportable quality of turmeric is actually minimising the demand in the external market.

7. In recent years, lower demand and swelling stock levels has reduced the scope of farmers in getting a remunerative price for turmeric.

Prospects

As turmeric has several vital properties and is being used both internally and externally in different respects, there is a bright

TABLE-VI: Country-wise Exports of Turmeric Oleoresins

(Qty in kg., Value in '000 Rs.)

Country	1994-95		1995-96		1996-97		1997-98	
	Qty.	Value	Qty.	Value	Qty.	Value	Qty.	Value
Australia	500	169	1169	431	1305	484	1710	1494
Canada	1010	432	1635	380	3763	2527	5681	6132
German FR	5581	2642	7018	3874	7265	2586	3264	3309
Switzerland	50	59	50	59	200	231	267	452
UK	22115	11527	16189	7935	18086	11339	28250	20515
USA	71582	38671	107907	52184	101310	55543	163085	79597
Total (including others)	120570	69388	161681	85542	154687	86733	157510	130745

Source: DG of CI and S, Calcutta.

future for this. In this regard the above said problems have to be solved without any delay by way of providing different incentives through a package of programmes. It should contain the following aspects:

(1) Research aspects should be initiated to reduce the crop duration period. Again studies has to be conducted on the physiology of the growth and development of rhizomes in turmeric.

(2) For increasing the productivity from unit area, a compatible inter or mixed cropping system should be developed. So as to solve the problem of diseases, integrated phytosanitary and curative measures should be evolved at the earliest to ensure better returns to the farmers.

(3) There is a need to evolve a well organized marketing system in all the producing states so as to minimise the fluctuations in the prices. Side by side more and more attention should be given to the extraction and marketing of oils and oleoresins.

(4) There is a need to evolve an integrated technology for the control of important pest and diseases and also to educate the farmers about the proper agrotechnology in scientific production of this crop. Again there is a need to create an efficient extension network.

(5) Need to organize scientific seminars, since, they provide a forum for all agencies concerned with production, protection, marketing and export.

(6) Efforts are needed to take the available technical know-how from the Lab to the common farmers which will help to reduce the yield gap and enhance production without any time lag.

(7) Processing and marketing research have to be taken up so that enough data is available to the policy makers to fix prices, formulate developmental and related policies.

(8) Research on bye-product utilization and product diversification should be undertaken.

(9) Proper post-harvest storage techniques has to be evolved for safeguarding the final crop produce from spoilage.

(10) A continuous surveillance of charges in technology, production and quality control of materials produced by other producing countries should be initiated for maintaining the quality of the Indian products in foreign trade.

(11) Agricultural Produce Market Act should be introduced and enforced wherever it is absent. In areas where this is already existing, necessary facilities for drying, grading and storage should be built up. The farmers should be educated for selling their produce in the regulated markets.

(12) Need to compile data on the production and inter-state movement of turmeric.

(13) Need to diversify the products and markets for boosting the exports.

As the demand for turmeric is increasing both internally and externally and its use as an anti-cancer agent would boost up further demand. Again popularization on the medicinal values of Indian organic turmeric in the US, Europe and Japan would boost demand further resulting in better prices to the growers. And as beauty aids, turmeric based products can be promoted as eco-friendly and allergy-free. So what is needed now is improving this sector and it's efficiency. Then only India can maintain her status in the world market for turmeric.

7

GARLIC

Garlic or *Lassan* needs no introduction since it has long been recognized all over the world as a valuable condiment for foods, and a popular remedy or medicine for various ailments and physiological disorders. It is a hardy bulbous perennial with narrow flat leaves and bears small white flowers and bulbs. Garlic grows under much the same conditions as the onion, except that it favours a richer soil and a higher elevation. A well-drained, moderately clayey loam is best suited for its cultivation. It requires cool moist period during growth and a relatively dry period during the maturing of crop. It takes 4-5 months to mature. In south India, it is rotated with chillies, ragi, potatoes, beans and maize. In the world, India and China are the major producers of garlic.

Area, Production and Yield of Garlic in India

Garlic has since long been cultivated practically throughout India as an important minor spice or condiment crop. The total area under garlic at present is about 98.5 thousand hectares and the production is 464 thousand tonnes while the yield is 4710 kg/hectare. The total area under garlic in the state of Madhya Pradesh is the highest one and it is 37.5 thousand hectares and the production is 170.4 thousand tonnes, this is followed by Gujarat, Rajasthan, U.P., Orissa, Maharashtra and Karnataka. The state-wise area, production and yield for the period 1995-96 to 1997-98 is given in Table-I.

TABLE-I: State-wise Area, Production and Yield of Garlic

(Area—'000 ha, Production—'000 tonnes, Yield—kg 'ha)

State	1995-96			1996-97			1997-98		
	Area	Production	Yield	Area	Production	Yield	Area	Production	Yield
Andhra Pradesh	0.7	1.2	1417	0.6	1.1	1833	0.6	1.0	1667
Bihar	2.7	4.0	1481	2.9	4.4	1517	3.0	4.0	1333
Gujarat	16.0	90.0	5625	16.1	102.0	6335	20.2	134.2	6644
Haryana	1.0	7.5	7500	0.6	7.4	12333	0.5	6.2	12400
J and K	0.4	0.3	750	Neg.	Neg.	—	Neg.	Neg.	—
Karnataka	4.0	3.1	775	4.7	3.7	787	4.6	3.5	760
Madhya Pradesh	41.4	188.9	4563	34.0	155.8	4582	37.5	170.4	4544
Maharashtra	6.7	45.9	6851	6.5	44.6	6862	6.5	43.6	6707
Orissa	18.5	52.1	2816	9.8	38.6	3938	6.7	26.5	3955
Punjab	0.9	11.2	12444	0.8	9.7	12125	0.9	10.0	11111
Rajasthan	14.1	46.1	3270	10.0	31.6	3160	9.9	30.1	3040
Tamil Nadu	1.1	6.7	6091	1.1	6.8	6182	1.0	5.9	5900
Uttar Pradesh	7.3	33.0	4521	7.2	32.2	4472	7.4	28.6	3865
All India	114.8	490.00	4268	94.3	437.9	4644	98.5	464.0	4710

Source: D of E and S, New Delhi.

As far as the total area under garlic in India is concerned, it has been increasing remarkably over the years, eventhough there appears to be fluctuations in it, as can be seen from Table-II. The maximum area under garlic was observed during the period 1995-96 that of 114.8 thousand hectares and the production was 490 thousand tonnes which is the highest one in recent years.

TABLE-II: Area and Production of Garlic in India

Year	Area in Hactares	Production in tonnes
1975-76	39800	140900
1980-81	59300	216000
1985-86	57700	189600
1989-90	83500	327000
1992-93	85500	355800
1993-94	76200	306000
1994-95	98900	403200
1995-96	114800	490000
1996-97	94300	437900
1997-98	98500	464000

Source: D of E and S, New Delhi.

Varieties

No distinct varieties of garlic are recognized in the northern part of India, however, in South India two distinct types, namely Fawari and Rajalle Gaddi with slightly bigger bulbs are grown and in other parts local names are given to the varieties which they grow in their regions. The speciality of Rajasthan garlic is that the flakes are big and the flavour is good.

The crop is ready for harvesting when the tops turn yellowish or brownish and show signs of drying up, usually about a month or so after the emergence of seed stalks. The bulbs are cured for three to four days in the shade before storing them in an ordinary room. Tops are removed before marketing the produce.

There are 4 varieties which are meant for trade, they are the 'medium' from Madhya Pradesh, 'VIP' quality, 'Bom' quality and the 'Super Bom' quality. The Super Bom fetches the highest price while the medium the lowest. Bangalore is the main exporting centre where both buying and selling of garlic takes place. From Bangalore, garlic is exported to Malaysia, Singapore and Sri Lanka.

Uses

Garlic is recognized all over the world as a valuable condiment for foods, and a popular remedy for various ailments and physiological disorders. The Unani and Ayurvedic systems say that, garlic is carminative and is a gastric stimulant, and thus aids in digestion and absorption of food. In modern allopathy, it is being used in a number of patented medicines and other preparations. The anti-bacterial action of garlic had been noticed from early days, its healing capacity and effectiveness against cholera having been recorded as early as eighteenth century.

As a condiment it is used practically all over the world for flavouring dishes. In USA nearly half of the entire output of fresh garlic is dehydrated and sold to food processors for use in mayonnaize products, salad dressing, tomato products and in several meat preparations. Further raw garlic is used in the manufacturing of garlic powder, garlic salt, garlic vinegar, garlic cheese and croutons garlicked potato chips, garlic bread etc. In Italy, Europe and particularly in Latin American countries, there appears to be a spectacular popularity for this. In recent years, there has been considerable demand from the food industries for garlic in India.

Garlic oil is an effective insecticide. Apart from this, it is also a valuable flavouring agent, for use in all kinds of meat preparations, soups, canned foods and sauces.

Garlic juice is used for various ailments of the stomach, as a ruberfacient in skin diseases and as ear-drops in ear-ache. The juice diluted with water can be used against duodenal ulcers. In Cambodia, the leaves are used in the treatment of asthma.

Export

India exports garlic, garlic paste and powder to South East Asian countries in larger volume, while for USA, Canada, EEC's it is in smaller quantum. Sri Lanka is the main consumer of Indian garlic followed by Malaysia, Singapore, Bangladesh, Kuwait and Saudi Arabia. During the period 1997-98, the total volume of garlic export was 3975 tonnes worth of 79706 thousands of rupees and this was 4889 tonnes valued at Rs. 79774 thousands in 1996-97. The trend of Indian export of Garlic as shown in Table-III indicates that there appeared ups and downs frequently in it. This type of trend has actually minimized the scope of the nation to earn the much needed foreign exchange. In 1998-99, the total expected export of garlic will be 8000 tonnes valued at Rs. 16 crores which is higher than the previous years exports.

TABLE-III: Export of Garlic from India

Year	Quantity (M.T.)	Value (Rs. '000)
1960-61	2030	1123
1965-66	795	713
1970-71	1634	2782
1975-76	931	2549
1980-81	7398	22824
1985-86	2619	13568
1990-91	4646	32772
1992-93	7442	71018
1993-94	2844	35489
1994-95	633	12287
1995-96	3936	9125
1996-97	4889	79774
1997-98	3975	79706
1998-99	5983	77615
1999-2000*	5000	150000

*Target

Source: DG of CI and S, Calcutta.

As far as our export trends are concerned, it clearly gives the idea that there is much to do for the improvement of this tangy spice. In this connection, the prevailing problems of this has to be solved. The problems are:

(1) Non-availability of high-yielding disease resistant varieties.

(2) Harvesting and post-harvesting problems like the prevalence of intermediaries in its marketing, non-availability of proper transport facilities and absence of storage facilities. Because of these problems, it is noted

that about 20-25 per cent of the fresh crop is wasted due to respiration, transpiration and micro biological spoilage.

(3) Under utilization of undersized or cull but healthy bulbs.

(4) Lack of publicity for garlic and its subsidiary products like garlic powder.

(5) Non-availability of proper training and education to the farmers with regard to pre- and post-harvesting aspects.

(6) Failure to find out alternative uses of garlic in a value added manner:

India once a market leader, is losing out in the overseas markets to China, the current market leader. So efforts are needed to tackle the above said problems and to supply qualitative garlic into the international market. In this regard, the following steps are useful.

(1) There is the need to identify new varieties with large bulbs of white colour, uniform shape and size fold and compact cloves, high yielding and resistance to diseases and pests. The new varieties should be thicker as that of China. It is because of this thickness Chinese garlic is in huge demand in the international market.

(2) Need to provide proper marketing facilities along with transportation and storage facilities to the farmers.

(3) As and when surplus crop appears, there is the need to channelise it into the production of garlic powder by improved techniques. This can minimize the wastes. This will also assist in regulating the market rates, especially in the glut season.

(4) All undersized or cull but healthy bulbs, which normally fetch lower price, could be conveniently utilized in the manufacture of garlic powder.

(5) Need to undertake demonstrations, publicity campaigns so as to popularize garlic powder within and outside the country.

(6) Garlic powder converted into suitable tablets or packed into capsules should find a ready market in India, in the face of costly synthetic tablets claiming anti-bacterial activity.

8

CORIANDER

Coriander seeds and fresh coriander leaves are well known spice, especially to the housewife. It is actually the housewife's secret of tasty dishes. It is a native of the Mediterranean region. Today coriander is commercially grown in India, Morocco, Russia, Hungary, Poland, Czechoslovakia, Rumania, Gautemala, Mexico, USA, Bangladesh and Pakistan.

In India, coriander is cultivated in practically all the states and it constitutes an important subsidiary crop in the black cotton soils of Deccan and South India and also in the rich silt loams of North India.

Uses

The fresh stem, leaves and fruits of coriander have a pleasant aromatic odour. The entire plant, when young, is used in preparing chutneys and sauces, and the leaves are used for flavouring curries and soups. The seeds are extensively employed as condiment in the preparation of curry powder, pickling spices, sausages and seasonings. They are used for flavouring pastry, cookies, buns, cakes and tobacco products. In countries like USA and Europe, it is employed for flavouring liquor, particularly Gin. The seed is used mainly after mild roasting. It is probably one of the first spices to be used by mankind, having been known as early as 5000 B.C., and is one of the important ingredients in the preparation of the food flavourings like bakery products, imitation flavour, pork, meat, fish and salads, soda and syrups, candy, preserves and liquors.

Since coriander seeds are considered to be carminative, diuretic, tonic, stomachic, antibilious, refrigerant and aphrodisiac they are used in the preparation of medicine. They are used chiefly to conceal the odour of the medicines and to correct the griping qualities of rhubarb and senna. These are also considered to lessen the intoxicating effects of spirituous liquors. An infusion of the seeds in combination with cardamom and caraway seeds is useful in flatulence, indigestion, vomiting and intestinal disorders. The volatile oil is used in the cocoa and chocolate industries. Oil of coriander seeds is a valuable ingredient in perfume, its soft, pleasant, slightly spicy note blends into scents of oriental character. A good quality of Oleoresin is used for flavouring beverages, pickles, sweets and other delicacies. The residue from distillation can be used as a good cattle feed as it is rich in protein and fat.

Area and Production of Coriander in India

In India coriander is mainly grown in Rajasthan, Madhya Pradesh, Andhra Pradesh, Tamil Nadu and Karnataka on a large scale. The total area under it is about 522 thousand hectares and the production is around 308 thousand tonnes in 1997-98. In terms of yield per hectare, it was 591 kgs over the years and the total area as well as production of this has been increasing, which is a positive sign of this sector. As far as state-wise production is concerned, Rajasthan covers an area of 232.6 thousand hectares under this and production is 215 thousand tonnes and the productivity is 923 kg/hectare. This is being followed by Madhya Pradesh and Tamil Nadu as can be seen from Table-I. In terms of all India area and production of coriander, it was 272 thousand hectares in 1970-71 while the production was 101 thousand tonnes which went up year after year and the maximum of 593 thousand hectares was observed in 1987-88 when the production was 29 thousand tonnes. However, from then onwards, the production under this went downwards upto 1996-97 but then onwards, it again moved in a positive direction as can be seen from Table-II.

TABLE-I: State-wise Area, Production and Yield of Coriander in India

(Area—'000 Ha, Production—'000 tonnes, Yield—kg.he)

State	1995-96			1996-97			1997-98		
	Area	Production	Yield	Area	Production	Yield	Area	Production	Yield
Andhra Pradesh	77.0	15.2	197	88.9	24.5	276	87.0	23	264
Bihar	2.4	1.1	458	2.9	1.4	483	3.0	1.4	467
Karnataka	14.8	2.5	169	15.3	2.6	170	17.0	2.9	170
Madhya Pradesh	113.4	40.0	353	134.0	50.8	379	135.6	50.6	369
Orissa	278.0	13.8	137	11.8	5.7	483	8.1	3.9	481
Rajasthan	137.7	116.4	845	157.1	154.8	985	232.6	214.9	923
Tamil Nadu	28.2	2.8	99	36.4	11.2	308	31.9	8.1	254
U.P.	6.3	4.3	683	6.3	4.5	714	6.4	3.9	609
Total	407.6	196.1	481	452.7	255.5	564	521.6	308.1	591

Source: D of E and S, New Delhi.

TABLE-II: Area and Production of Coriander in India

Year	Area (in hectares)	Production (in tonnes)
1970-71	272300	101200
1975-76	242400	93700
1980-81	280100	110800
1985-86	382100	117100
1989-90	332700	143100
1992-93	402600	191200
1993-94	464800	203700
1994-95	430400	193000
1995-96	407600	196100
1996-97	452700	255500
1997-98	521600	308100

Source: D of E and S, New Delhi.

Several varieties are made available for trade, the important among them are Rajasthan Green, Guntkal varieties which fetch higher prices. Marketing channel for coriander consists of local merchants, intermediaries, secondary markets etc., they grade it like low quality, medium best and superior. The main domestic marketing centres are Ramaganjamandi in Rajasthan, Chennai city in Tamil Nadu, Bangalore in Karnataka and Mumbai in Maharashtra.

Exports

India exports coriander mainly to Malaysia, Singapore, UAE, UK, Sri Lanka, USA in larger volume while a smaller quantum is being exported to Kuwait, Japan and Nepal (Table-III on page). As far as the export of this is concerned, it has been fluctuating over the years, which is not a good sign as far as the development of the country is concerned. The total volume of coriander export in 1960-61 was 1859 metric tonnes and the value of it was Rs. 2062 thousands, which came down 142 metric tonnes in 1965-66 and went upto 515 metric tonnes in 1969-70 and from then onwards, it slowly went upward as can be seen from Table-IV on page . Since 1992-93, it has been showing several ups and downs in the export front. However, coriander exports are likely to create new records in 1998-99, touching around 25,000 tonnes, because of bumper crop and low prices.

The above aspects clearly indicate that our performance in the international market for coriander is not an encouraging one. This is so because of inherent problems of this sector. They are:

(1) The Indian coriander being rather poor in oil content, which is around 0.8 per cent, whereas the fruits from European countries are usually rich in oil, in Norway it is as much as 1.4-1.7 per cent and is upto 2 per cent in Russia. The low oil content of Indian coriander is there due to the loss of a portion of the volatile oil during the drying of fruits, too much splitting of fruits and faulty harvesting procedure practiced by the growers.

(2) Indian coriander seeds are low in essential oil and are not uniform in colour as well as grade, so it is not generally preferred for the distillation of essential oil by Western countries.

TABLE-III: Country-wise Exports of Coriander

(Quantity—Tonnes, Value—'000 Rs.)

Country	1995-96		1996-97		1997-98	
	Qty.	Value	Qty.	Value	Qty.	Value
Singapore	2592	49374	2995	78057	3774	103382
UK	634	13684	597	21027	1196	44757
Malaysia	3165	58575	3631	88265	4688	130594
Nepal	1	34	3	112	78	1664
Kuwait	82	2241	153	5250	113	5149
Japan	30	693	20	450	77	2307
UAE	1623	29791	1337	32937	3214	81387
Sri Lanka	463	6713	1400	24889	875	17379
USA	220	7824	159	6774	376	14075
Total (including others)	11541	224334	12574	313658	20901	591985

Source: DG of C and S, Calcutta.

TABLE-IV: Export of Coriander from India

Year	Volume (M.T.)	Value (Rs. '000)
1960-61	1859	2062
1965-66	142	313
1970-71	393	974
1975-76	755	3459
1980-81	2161	12556
1985-86	1864	16006
1990-91	3236	38015
1992-93	13737	210258
1993-94	13552	210351
1994-95	10702	179384
1995-96	11541	224334
1996-97	12574	313658
1997-98	20901	591985
1998-99	26826	573517
1999-2000*	20000	500000

*Target

Source: DG of CI and S, Calcutta.

(3) Lack of proper marketing facilities nearer to the production centre.

(4) Non-availability of high-yielding varieties of qualitative seeds to the farmers.

(5) The price of Indian coriander which enters into the external market is too high when compared to the competitors.

(6) Adulteration is a basic problem. The whole coriander often contains stalks, dirt, fenugreek, some cereals and other extraneous matter, which shows that proper grading and standardization is absent.

Since, there appears to be a potential market for coriander seeds in the foreign countries, there is an urgent need to overcome the prevailing hurdles of this sector. Again comparing with our production, the volume of export is also not a healthy one. Hence there is the need to adopt a long-term strategy for the benefit of this sector. The urgent need of the hour is to evolve new suitable high-yielding, firm like Russian varieties which are not easily breakable and disease-resistant varieties rich in essential oil adoptable to the various agro-climatic regions of the country. This will ultimately increase the country's production so as to meet the internal and external demands. Recently introduced improved varieties like CO-1, CO-2, CO-3, CS 287, RCr-20, RCr-41, Sadhana, Swathi, Sindhu, Pant Dhania, Hisar-Anand etc., should be made available to the farmers to increase the production.

By an increase in our production along with proper grading or maintaining seed quality, it is possible to enlarge our exports. Along with this there is also need to reduce the cost of production which in turn keeps our price of coriander at par with that of the competitors in the International Market. Again there is the need to reduce the export cess and to provide other incentives to encourage export at a time when huge exportable surplus is available.

For promoting our exports, there is the need to popularize the various uses of coriander in the foreign market through demonstrations, publication etc. As practically all the parts are most useful and the demand for this spice, both in the domestic and external market is ever increasing, co-operation and co-ordination between the government, research institutions and farmers is very essential to achieve these objectives. An extension of area under coriander cultivation along with education to the farmers all over India is basically needed which will be useful to realise the expectations in this sector. Again, the spices Board should promote this crop along with cardamom and pepper.

9

CELERY

Celery seed is the dried ripe fruit of the umbelliferous herb. It contains a small seed, united or separated pericarp, some with stalk ends, brown in colour and somewhat bitter in taste. The epicarp is intercepted with oil ducks. It is widely used as a spice.

The native habitat of celery extends from Sweden to Egypt, Algeria and Ethiopia, and in Asia, to India, Caucasus and Baluchistan. Celery is cultivated both for salad and seed-raising in the north-west Himalayas, Punjab, U.P. in India and the other countries where this is grown are France and USA. It is in great demand both in India and abroad.

Celery is a moisture-loving plant, requiring a cool climate. In colder climates and on the hills, it is biennial crop and produces seeds only in the second year, while in the plains, it becomes an annual crop and produces seeds in the very first year.

Uses

The dried ripe fruits are used as spice leaves and stalks are used as salads in soups and as a pre-dinner appetizer, leaves are more nutritions than stalk, particularly from the viewpoint of protein, Vitamins A and C. The seeds are stimulant and tonic is used in Asthma and liver diseases. As a domestic medicine, seeds are used as nervine sedative and tonic. These seeds are the main ingredient of celery tonic. The oil of celery is used as an anti-spasmodic and nerve stimulant. It has been successfully employed in rheumatoid arthritis and probably acts as an intestinal antiseptic. Its seed is used as a bird food also.

The industrial use of celery seed oil is as a fixative, as an important ingredient in novel perfumes, in medicine and in the flavouring of different kinds of foods—canned soups, meats, sausages and particularly in the flavourising of the popular celery salts, tonics and culinary sauces. Celery seed oil is one of the most valuable flavouring agents as it imports a warm, aromatic and pleasing flavour to food products.

The total production of celery seed in 1970-71 was about 3,500 tonnes in India and it went up to 4,000 tonnes in 1980-81 however in 1990-91 it came down to 3000 tonnes, but then it went up to above 5000 tonnes in the forthcoming years. These ups and downs in the production were observed mainly because of the natural conditions and lack of interest from the side of growers.

Exports

India exports celery seed to USA, Canada, France, Japan, Australia, UK etc., (Table-I) USA is the main market for Indian Celery seeds. In 1997-98 USA imported 2247 tonnes of celery seeds from India worth Rs. 51,584 thousand. The total volume of our export of celery seeds in 1997-98 was 3317 tonnes worth Rs. 79,919 thousand as can be seen from Table-II. In 1960-61 we, exported 1,398 tonnes of celery seeds to the International Market, and this has been increasing over the years. These figures give an impression that there is vast scope for increasing the export of this seed in the coming years.

So as to expand the exports, there is the need to promote the total production of this fruit in India. This calls for certain preplanned steps. First of all, interest has to be created among the growers to cultivate it in a efficient manner. In this regard, proper training is to be provided. Apart from this, training is also needed to improve the quality of this product, since celery seed is subject to adulteration by addition of exhausted or spent seeds, excess stems, chaff etc. The ground celery is sometimes adulterated with

TABLE I: Country-wise Export of Celery Seed from India

(Qty. tonnes, Value '000 Rs.)

Country	1995-96		1996-97		1997-98	
	Qnty.	Value	Qnty.	Value	Qnty.	Value
Australia	26	703	25	609	53	1810
Belgium	25	798	18	403	36	969
Canada	121	2845	135	341	153	3834
France	92	2444	28	563	128	3183
GFR	153	3923	110	2806	143	3791
Japan	113	2693	83	1892	93	2292
Netherland	72	1589	90	2116	68	1808
Singapore	40	993	21	703	34	1002
UK	128	2956	116	3130	111	2909
USA	1753	39636	2924	59902	2247	51584
Total (including others)	2678	62512	3780	80176	3317	79919

Source: DG of CI and S, Calcutta.

farinaceous products, linseed meal, worthless vegetable seeds or
weed seeds etc., all of these reduce the demand in the external
market. Again there is vast scope for its utilization in different
types of industries both domestically and externally. Hence,
concentrated efforts are needed to increase the production and
exports of celery seeds in the near future.

TABLE-II: Export of Celery from India

(Qnty.—tonnes, Value '000 Rs.)

Year	Quantity	Value
1960-61	1398	2079
1965-66	2150	3871
1970-71	3138	20202
1975-76	2243	11472
1980-81	3198	15635
1985- 86	2625	30682
1990-91	2598	33527
1992-93	3437	45769
1993-94	4130	66262
1994-95	2601	77728
1995-96	2678	62512
1996- 97	3780	80176
1997-98	3317	79919
1998-99	4882	122042
1999-2000*	3600	90000

*Target

Source: DG of CI and S, Calcutta.

10

CLOVE

Clove the "Royal Spice" is the dried unopened mature flower buds. Clove is the second most important spice of the world, as judged from the world trade, being next only to black pepper. Clove was in use as early as in the third century B.C. in China, was well known to the Romans and reached Europe during the Middle Ages. Today it is grown in many tropical countries. It is interesting to note Tanzania alone supplies about 90 per cent of the total world market for clove. The biggest clove producing region in the world today are Zanzibar, Pemba, Madagascar and Indonesia. It is also grown in Malaysia, Sri Lanka, India and Haiti in small quantities.

In India, clove was introduced in 1800 A.D. by the East India Company. The company's spice garden in Courtallam in Tamil Nadu was then established to cultivate clove and nutmeg as the principal spice crops. Induced by the success of its cultivation of clove, it was extended during the period after 1850 to Nilgiris in Tamil Nadu, southern regions of Travancore and also to Cochin on the slopes of Western Ghats.

In India, the cultivation of clove is mainly confined to South India viz. Karnataka, Kerala, Tamil Nadu, Andaman and Nicobar Islands. The total area under this crop is about 2500 hectares and the production is around 2300 tonnes. The total area under this crop in 1950-51 was about 1399 hectares and the production was 1000 tonnes. The area under this crop went up steadily and in 1989-90, it was 1943 hectares and the production was 1800 tonnes. Still it went up in the following years and in 1995-96 it reached to

2500 hectares in India. However, since then area under this crop has not moved further. Of the total area under this crop, the dominance of Karnataka can be seen, which constitutes an area of about 877 hectares followed by Tamil Nadu and Kerala 810 and 750 hectares respectively in 1995-96. In Karnataka, the area under this crop was only 75 hectares in 1992-93, however, from then, it went up sharply.

Clove is strictly a tropical plant which requires a warm humid climate. Although there has been a general belief that clove requires proximity to sea for the proper development, cropping experience in India has shown that the trees do well in the hinterland conditions too. In fact clove tree growing in submontane regions has been found to perform better than those in other areas. In India clove has developed well in the open sandy loams and the laterite soils of southern region. However, the best growth is seen in black loams of the semiforest region.

The clove of commerce is the air-dried unopened flower-bud obtained from a handsome, medium sized evergreen straight trunked tree that grows to a height of 10-12 metres.

Harvesting

Clove clusters are picked by hand when the buds have reached their full size and most of them develop a pronounced pink flush. From these clusters, cloves are separated from the stems consisting of peduncles and pedicel. Then it is dried. Dry cloves weigh one-third weight of green, freshly harvested cloves. Quality of the dried spice is influenced by a number of factors which include the care taken in harvesting, drying and storing.

Uses

Clove is very aromatic, has a fine flavour and imparts warming qualities. It is used as a culinary spice as the flavour blends well both sweet and savoury dishes. Cloves, both whole and ground, are used in baked goods, cakes, confectionery, chocolate, puddings, desserts, sweet syrups, preserves etc. It is also used for flavouring curries, gravies, pickles, ketchup and sauces, spice mixtures. It is highly valued in medicine as carminative, aromatic and stimulant. Clove has stimulating properties and is one of the ingredients of betelnut chew. In Java it is used for making a special brand of cigarette for smoking. The essential oil, which is obtained by distilling clove with water or steam has even more uses, which is used in medicine as an aid to digestion and for its antiseptic and antibiotic properties in tooth-ache. It is an ingredient of many toothpastes and mouth washes. The oil has many industrial applications and is extensively used in perfumes in scenting soaps, as a cleaning agent etc. It is also used for the flavouring of oral preparations and chewing gums.

Imports

India along with USA is the largest importer of cloves in the world. The domestic production is insufficient to meet requirements, so it is met through the imports. India imports both exhausted and unexhausted cloves along with clove leaf oil and several aromatic chemicals derived from its oil.

India import clove from Tanzania, Singapore, Malaysia, Sri Lanka, Indonesia and China. During 1971-72, India imported 1.42 lakh kg. of exhausted cloves and over 8000 kg. unexhausted cloves. In 1992-93 it was 2924 tonnes valued at Rs. 95008 thousand and in 1997-98 it went upto 6995 tonnes and the value was Rs. 271070 thousand of rupees. As far as country-wise imports are concerned, the maximum comes from Tanzania followed by Singapore and Sri Lanka (Table-I). With regard to import of clove oil, India imported

TABLE-I: Country-wise Import of Clove into India

(*Qnty. in tonnes, Value '000 Rs.*)

Country	1996-97		1997-98	
	Qnty.	Value	Qnty.	Value
China	31	1012	33	1176
Indonesia	—	—	9	135
Malaysia	685	21197	—	—
Singapore	1236	39493	1186	19799
Tanzania	2029	65538	2349	87076
Sri Lanka	329	11988	520	18966
Total (including others)	5254	169423	6995	271070

Source: DG of CI and S, Calcutta.

143405 kgs. of it in 1992-93 valued at Rs. 12970 thousand and it went
upto 214230 kgs. in 1996-97 but the demand came down to 169520
kgs in 1997-98 valued at Rs. 24254 thousand. (Table-II)

TABLE-II: Import of Clove Oil into India

Year	Qty., (kg.)	Value ('000 Rs.)
1992-93	143405	12970
1993-94	195354	14870
1994-95	101113	8171
1995-96	184178	73681
1996-97	214230	30955
1997-98	169520	24254

Source: DG of CI and S, Calcutta.

All these aspects clearly shows that India is spending large
volume of the much needed foreign exchange for the import of
clove and its bi-products, eventhough it is in a position to produce
the required volume internally, So, there is the need to extend the
area under this highly useful and valuable crop in order to
minimize the drain of valuable foreign exchange.

Although it has been under cultivation in India for over two
centuries, its development has not been made up. The major
reasons for this are its long re-bearing age, lack of scientific
knowledge regarding cultivation, harvesting, marketing and other
problems. The farmers are worried about picking up flowers from
the long trees and also finding a proper market for the same.

To attain self-sufficiency, it is noted that atleast 15,000 acres
should be brought under this crop. This calls for a planned strategy,
which can solve the sectors problem and can also be useful to
increase the production. The experience in Kerala and Karnataka
shows that clove can be conveniently grown mixed with other
commercial crops like arecanut, coconut etc. Hence government
should take steps in this regard so as to popularize this crop in
India.

11

CUMIN

Cumin comprises the dried yellowish to greyish brown seeds of a small slender annual herb of the coriander family, believed to be the native of Egypt and Syria, Turkestan and the Eastern Mediterranean region. It grows to a height of 30-45 cms and produces a stem with many branches bearing long, finely divided, deep green leaves and small flowers, white or rose colour, borne in umbels. This aromatic seed like fruit, commonly known as 'seed' is elongated, oval and light yellowish brown in colour, somewhat similar to caraway seed but slightly longer. The odour is peculiar strong and heavy, pleasant to some and rather disagreeable to others, while flavour is warm, slightly bitter and somewhat disagreeable.

Cumin is one of the oldest spices, known since Biilical times. Today, this is grown extensively in Iran, India, Morocco, China, Southern Russia, Indonesia, Japan and Turkey. Iran is the major producer as well as exporter of the cumin seed which is popularly known as 'green cumin'.

Area, Production and Yield of Cumin in India

In India, cumin is cultivated in almost all the states except Assam, Kerala and West Bengal. However, the major producers are Rajasthan, Gujarat and U.P. The total area under Cumin in India at present is around 2,90,000 hectares and the production is about 1,16,265 tonnes. The area under this has been fluctuating over the years in our country. In Rajasthan, it is cultivated in an area of around 1,67,900 hectares and the production is about 57,135 tonnes while Gujarat covers an area of 1,20,900 hectares and the production is 58,200 tonnes which is higher than Rajasthan eventhough the total area under cumin is less. Again in terms of yield, per hectare, in Gujarat, it is 481 kgs whereas it is only 340 kgs. in Rajasthan (Table-I). In 1975-76, the total area under this crop was about 86,880 hectares and the production was 28,170 tonnes. Then, the area under this increased at a faster rate and has reached 1,80,549 hectares in 1978-79 and the production was about 79116 tonnes. However, it came downwards from then onwards slowly and was 80,743 hectares in 1986-87 and again it went upward and in 1993-94 a maximum area under this was noted at 4,20,782 hectares, but once again it declined in the future. (Table-II).

Uses

Cumin seeds have an aromatic odour and a spicy and somewhat bitter taste. They are largely used as condiment and form an essential ingredient in all mixed spices and curry powders for flavouring soups, pickles and for seasoning breads and cakes. These are considered as stimulant, carminative, stomachic, astringent and are useful in diarrhoea and dyspepsis. Cumin seeds are now highly used in veternay medicine. The volatile cumin seed oil is used in many types of flavouring compounds, especially in curries and culinary preparations of oriental character. The oil is also used in perfumery and for flavouring liquors and cordials.

Exports

India exports cumin in American zone, Australia and Oceanic, EEC, East and West Asian zone and also to the African zone. As far as country-wise export is concerned, USA is the major market followed by Nepal, Singapore, UK and others in the list. The total volume of cumin export to USA in 1997-98 was 2615 tonnes and the value was Rs. 152597 thousand, while for Nepal it was 1817 tonnes and Singapore it was 1727 tonnes during the same period. (Table-III). As a whole, in recent years, our exports of cumin have

TABLE-I: State-wise Area, Production and Yield of Cumin in important Producing States in India

(Area '000 he, Production in '000 tonnes, Yield kg/he)

States	1995-96			1996-97			1997-98		
	Area	Production	Yield	Area	Production	Yield	Area	Production	yield
Rajasthan	125809	34033	286	198128	65714	332	167911	57135	340
Gujarat	94500	39200	415	108900	51400	472	120900	58200	481
U.P.	17	753	442	18	797	442	21	930	442
Total	220326	75241	341	307045	117911	382	288832	116265	403

Source: Spices Board, Cochin.

TABLE-II: Area and Production of Cumin in India

Year	Area[+] (in hectares)	Production[+] (in tonnes)
1975-76	86880	28170
1980-81	165305	89155
1985-86	73629	35475
1989-90	150493	46874
1992-93	313920	135200
1993-94	420782	166535
1994-95	277327	98669
1995-96	220326	75241
1996-97	307045	117911
1997-98	288832	116265
1999-2000[++]	10000	600000

Target[++]

+ Include figures for major producing states only.

Source: State Agricultural departments.

TABLE-III: Country-wise Export of Cumin Seed from India

(Qnty. in tonnes, Value '000 Rs.)

Country	1995-96		1996-97		1997-98	
	Qnty.	Value	Qnty.	Value	Qnty.	Value
Canada	80	3991	121	6892	103	5664
USA	302	16343	895	51638	2615	152597
Singapore	173	8455	733	37646	1717	79269
Japan	398	23204	538	35908	535	34745
UK	390	18834	349	20011	798	44367
Bangladesh	—	—	320	14310	188	6664
S. Arabia	11	267	10	560	130	6082
Morocco	—	—	14	804	114	6543
Nepal	1555	71093	1451	84842	1817	89092
Malaysia	38	1721	320	15138	207	30814
Total (including others)	3870	173932	6375	343780	16195	809343

Source: DG of CI and S, Calcutta.

been increasing (Table-IV), which gives the impression that there is a vast scope for our exports in the years to come.

TABLE-IV: Export of Cumin from India

Year	Quantity (M.T.)	Value (Rs. '000)
1960-61	1201	2011
1965-66	3516	10268
1970-71	2363	8262
1975-76	2493	20154
1980-81	8778	97426
1985-86	1061	16570
1990-91	1025	29679
1992-93	2620	143887
1993-94	3225	163034
1994-95	5618	244965
1995-96	3870	173932
1996-97	6375	343780
1997-98	16195	809343

Source: DG of CI and S, Calcutta.

India, being a major producer of cumin seed in the world, can definitely lead other producing countries, provided high-yielding and disease-resistant varieties, rich in essential oil content are evolved to suit various agro-climatic regions of the country. Varieties like S-404, MC-43, RZ-19, Guj Cumin-1, Guj CUM-2 has to be supplied to the farmers, then only production and productivity can be increased. By increasing our production and by maintaining the seed quality, we can bring the price of cumin at par with those of the international competitors. In Holland and Switzerland, cumin is used to season some kind of cheese, while in France and Germany, it is used to flavour bread, cakes and pastries. Cumin makes an excellent seasoning of soups and stews, for which it is widely used throughout Latin America. It is also employed in native dishes of Central and South America. Because of this growing popularity of cumin, the demand for Indian cumin seed is always on the increase as can be seen from the export trends over the years. So as to maintain this happy trend, there is the ultimate need to increase the production in the country, further, by popularising the uses of cumin, India's export could be possibly increased substantially.

12

FENNEL

Fennel, the dried, ripe fruit of cultivated varieties of Foeniculum vulgare Mill., belongs to family Umbelliferae, which is a biennial or perennial aromatic, stout, glabrous herb with a height of 1.5-1.8 metres, cultivated in Mediterranean countries, in Romania, and in India. The seed of this is small, ablong, ellipsoidal or cylindrical having greenish yellow or yellowish brown colour and it possesses an agreeable, aromatic and sweet aroma resembling aniseed.

Fennel is cultivated in India mostly as a garden or homeyard crop throughout the country at all altitudes upto 1,825 metres. It requires a fairly mild climate and is cultivated as a cold-weather crop in northern parts of India. Apart from India, Russia, Hungary, Germany, France, Italy, Japan and others are also growing fennel in the world.

Uses

Fennel plant is pleasantly aromatic and is used as a pot-herb. The leaves are used in fish sauce and for garnishing; leaf stalks are used in salad. Thickened leaf stalks of Florence fennel are blanched and used as vegetable. The leaves are possessing diuretic properties and the roots are regarded as purgative and are having an aromatic odour and taste. In India and in neighbouring countries like Pakistan and Sri Lanka, fennel is used as a masticatory or for chewing alone or in 'paans'. It is also used for flavouring soups, meat-dishes and sauces, bread-rolls, pastries and confectionery, liquors, and in the manufacture of pickles.

TABLE-I : State-wise Area and Production of Fennel Seed in the Important Producing State of India

(Area in Hectares, Production in tonnes)

State	1995-96		1996-97		1997-98	
	Area	*Production*	*Area*	*Production*	*Area*	*Production*
Rajasthan	2721	1997	6407	5180	6307	5281
Gujarat	10100	13700	18700	23200	20500	31300
U.P.	645	307	632	315	632	315
Total (including others)	13466	15004	25739	28695	27439	36896

Source: Spices Board, Cochin.

The fruits are aromatic, stimulant and carminative. They are official in the pharmacopoeias of all countries and are considered useful in diseases of the chest, spleen and kidney. They are employed as a corrective for less pleasant drugs, particularly senna and rhubarb. During the Second World War, fennel oil was utilised in place of anise oil as a source of anethole. It is useful in infantile colie and flatulence.

Area and Production of Fennel in India

In India fennel is grown in Gujarat on an area of 20,500 hectares and the production is around 31,300 tonnes and it occupies the first place both in terms of area under this and also in production followed by Rajasthan, U.P., Maharashtra and Karnataka. The total area under fennel in 1975-76 was 14,566 hectares and it upto 27,908 hectares in 1983-84, however, it came down to 5,740 hectares in 1987-88, and from then onwards it went up. On the other hand in terms of production, it was 17,443 tonnes in 1975-76 and according to area, the production also fluctuated over these years. In 1997-98, the total area under this crop was 27,439 hectares and production was 36896 tonnes (Table-I and II).

TABLE-II: Area and Production of Fennel in India

Year	Area + (in hectares)	Production + (in Tonnes)
1975-76	14566	17443
1980-81	11785	13239
1985-86	9959	10401
1989-90	11700	12150
1992-93	19568	22691
1993-94	14908	15173
1994-95	25244	29349
1995-96	13466	15004
1996-97	25739	28695
1997-98	27439	36896

+ *Include figures for major producing states only.*
Source: State Agricultural Departments.

TABLE-III: Country-wise Exports of Fennel Seeds from India

(*Qnty. in tonnes, Value '000 Rs.*)

Country	1995-96		1996-97		1997-98	
	Qnty.	*Value*	*Qnty.*	*Value*	*Qnty.*	*Value*
Canada	81	2409	83	4185	116	5003
Kuwait	41	1625	14	780	43	1470
Malaysia	113	3399	447	15131	806	20524
UAE	316	7609	442	13498	1338	33639
Singapore	102	2984	905	26112	778	19964
S. Arabia	219	8134	314	13580	350	11480
USA	684	19774	659	27426	1710	59859
UK	246	8416	296	13519	337	13049
Total (including others)	2595	75173	2594	74887	12159	353999

Source: DG of CI and S, Calcutta.

Exports

USA is the major market for Indian fennel followed by UAE,
Singapore, Malaysia and others. India exported 12,159 tonnes of
fennel during 1997-98 and its value was Rs. 3,53,999 thousand. As
compared to the previous years, this was considered to be the
highest one over the years as can be seen from Table-III on page
and IV. The whole trend in our exports indicate that their appeared
several ups and downs. The main reason for this is fluctuating
global demand, lower domestic output coupled with high rates in
local markets, making Indian fennel uncompetitive in the overseas
market. The main competitors are Syria and Pakistan. However,
Indian fennel is of superior quality as compared to fennel from
competitors. This type of tendency is actually retarding our scope
to earn the much needed forex. So as to gain the real benefits from
external market, there is the need to overcome the prevailing
hurdles of this sector.

TABLE-IV: Export of Fennel from India

(Qty. in tonnes, Value Rs. '000)

Year	Quantity	Value
1960-61	1389	1516
1965-66	1487	2265
1970-71	795	2726
1975-76	615	5571
1980-81	1416	10724
1985-86	1325	10631
1990-91	1153	19922
1992-93	3007	70307
1993-94	2637	64218
1994-95	2029	58156
1995-96	2595	75173
1996-97	2594	74887
1997-98	12159	353999
1998-99	8007	240570
1999-2000*	4900	220000

*Target
Source: DG of CI and S, Calcutta.

Problems and Prospects

This basic need of this sector is the selection of high essential oil yielding and high seed yielding varieties. The Agricultural Universities and Research stations should identify such a variety which could meet the requirements of growers, and will be useful to enhance the production and also that of our exports. The percentage of oil being the lowest in fruits of Indian origin as compared to the Eastern Europe varieties. In this regard, efforts are needed to identify such type of varieties as that in Eastern Europe. Recently identified improved varieties like S. 7-9, PF-95, Guj-Fennel-1, Guj-Fennel-2, CO-1, Rajendra Saurabh should be supplied to the farmers without any delay.

As far as the quality of fennel which is exported is concerned, it is the lowest one. This is so because of poor harvesting and storing practices followed by the growers. Again, these often contain sand, stem tissue, stalks and other umbelliferous seeds. They are sometimes adulterated with exhausted or partially exhausted fruits or with undeveloped or mould-attacked fruits. There is also some confusion between the seeds of fennel and those of anise. Hence, what is needed is to ensure better post-harvesting practices. In this regard, the concerned authorities should provide proper education and training to the growers, then only Indian fennel can gain an increased demand in the global market.

Apart from all these, the most worrying aspect is that a large portion of the domestic production is consumed within the economy. The production and exports in terms of volume says much about this. Of the total production, India used to export only 10-30 per cent over the years. So the need of the hour is an increase in the production and productivity with the help of high-yielding varieties along with modern methods of pre- and post-harvesting activities and by providing incentives to the farmers since in recent years farmers switched over to other more lucrative crops because of the low returns in this fennel crop.

13

FENUGREEK

Fenugreek, an important spice, is grown throughout India. It is the dried ripe fruit of an annual herb, native of South Eastern Europe and West Asia and at present cultivated in India, Argentina, Egypt and Mediterranean countries. The seed is small and yellowish brown in colour. It has a pleasantly bitter taste and a peculiar odour and flavour of its own. The seed is produced as a spice, as a vegetable for human consumption, as forage for cattle and to some extent for medicinal purposes.

The fresh tender pods, leaves and shoots which are rich in iron, calcium, protein, Vitamin A and Vitamin C are eaten as curried vegetable since ancient times in India, Egypt etc. As a spice fenugreek adds to the nutritive value and flavour of foods. For this, fenugreek is of considerable importance in those countries in the Middle and Far East where meatless diets are customary for cultural and religious reasons. In Egypt and Ethiopia, fenugreek is a popular ingredient of bread, known to the Arabs as 'hulba', and in Ethiopia going by the Amharic name 'abish'. In Greece, the seeds, boiled or raw, are eaten with honey. In the United States, it is used in the manufacture of chutneys and in various spice blends, but its most important culinary use is as a source of fenugreek extract, the principal flavouring ingredient of imitation maple syrup. It is used in recipes like Hearty Vegetable Bean soup and

Fenugreek Beef stew. It is mainly of interest as one of the principal odorous constituents of curry powder.

Medical papyri from ancient Egyptian tombs reveal that it was used both to reduce fevers and also as a food. The belief of the ancients, it stimulates the digestive process as well as the metabolism in general. The seeds are used in colic flatulence, dysentery, diarrhoea, dyspepsia with loss of appetite, chronic cough, dropsy, enlargement of liver and spleen, rickets, gout and diabetes. The seeds are used as carminative, tonic, aphrodisiac; and infusion given to small pox patients as a cooling drink; toasted and then infused, used in dysentery. It is also used in sweets served to ladies during the post-natal period.

In the Middle Ages, fenugreek was recommended as a cure for baldness in men. In Java, it is used in hair tonic preparations and as a cosmetic. The powder made from the seeds is used in the Far East as a yellowish dye. In recent years, its oil has attracted the interest of the perfume industry.

Area and Production of Fenugreek in India

The total area under fenugreek in India at present is around 50000 hectares and the production is about 60,000 tonnes. Over the years, the total area under this is more or less remaining stable and in terms of production also it remains the same. Rajasthan possesses an area of 33,000 hectares and its production is about 31000 tonnes and stands first in the list of producers followed by Gujarat where 7100 hectares of land is used for the cultivation of fenugreek and the production is about 8100 tonnes, followed by states like Haryana and U.P. as can be seen from Table-I. Table-II giving figures for area under the production of fenugreek in India which includes major producing states.

TABLE-I: State-wise Area and Production of Fenugreek Seed in the Important Producing States in India

(Area—Ha, Production—Tonnes)

State	1995-96		1996-97		1997-98	
	Area	Production	Area	Production	Area	Production
Rajasthan	32905	40759	32887	43434	33051	31110
Uttar Pradesh	530	339	534	307	539	303
Gujarat	5600	6400	NA	NA	NA	NA
Haryana	NA	NA	1775	2800	NA	NA
Total	39035	47494	35196	45841	33590	31413

Source: State Departments.

TABLE-II: Area and Production of Fenugreek in India

Year	Area + (in hectares)	Production + (in Tonnes)
1975-76	31164	43473
1980-81	38478	52636
1985-86	30956	31953
1989-90	37635	38806
1992-93	30922	34686
1993-94	35778	37872
1994-95	46300	57146
1995-96	46300	57146
1995-96	39035	47494
1996-97	35196	45841
1997-98	33590	31413

+ *Include figures for major producing states only.*
Source: State Agricultural Departments.

Exports

Considering the total volume of production of fenugreek in India, the volume of its export is just 10 per cent of the total. The total volume of our export in 1997-98 was 5570 tonnes valued at Rs. 93,309 thousand, which was Rs. 1,86,720 thousand in 1995-96 with an export of 15138 tonnes. Over the years, the volume of fenugreek export from India has been fluctuating as can be seen from Table-III. The markets for Indian fenugreek seeds are YAR, Japan, UK, S.Arabia, Sri Lanka, USA etc. In terms of the volume of import of fenugreek seeds from India to these countries, it shows several ups and downs over the years. The comparative figures given in Table-IV (on page) support this view. This type of condition appears mainly because of the poor production pattern, excess domestic consumption, absence of pre- and post-harvesting, up to date technology, and lack of initiative from the departmental side to promote and protect this sector.

TABLE-III: Export of Fenugreek from India

(*Qty. in tonnes, Value Rs. '000*)

Year	Qty.	Value
1960-61	799	693
1965-66	981	991
1970-71	1042	1458
1975-76	1541	3982
1980-81	4470	17344
1985-86	2394	9890
1990-91	3449	28396
1992-93	5253	56964
1993-94	4934	72141
1994-95	7956	122492
1995-96	15138	186720
1996-97	8891	120457
1997-98	5570	93309
1998-99	12231	224146
1999-2000*	9000	180000

*Target

Source: DG of CI and S, Calcutta.

Prospects

India being a major producer of fenugreek can overtake other countries which are producing and exporting this to international market, provided we evolve high-yielding and disease-resistant varieties suited to our various agro-climatic conditions. Recently identified improved varieties like CO1, CO2, Rajendra Kanti, RMt-1, LamSel-1, Hisar Sonali should be made available to the farmers so as to increase the productivity. In this regard, the need of the hour is to carry out scientific and technological research on this important minor spice of our country. Again, more research is needed to improve the roasting aspects, since better methods of roasting of seeds significantly improve the flavour.

As fenugreek seeds contain many substances like protein, starch, sugars, mucilage, mineral matter, volatile oil, vitamins and enzymes, efforts are needed to improve the production of this

TABLE-IV: Country-wise Export of Fenugreek Seed from India

(Qty.—Tonnes, Value—'000 Rs.)

Country	1995-96		1996-97		1997-98	
	Qnty.	Value	Qnty.	Value	Qnty.	Value
Australia	0.1	17	6	174	0.82	23
Japan	402	5650	800	11005	596	9320
Korean RP	250	3216	300	3708	250	3867
Kuwait	118	2166	66	1175	36	727
Nepal	2	20	24	410	44	590
Sri Lanka	1301	14180	1103	12577	291	3539
S. Arabia	582	7590	399	4961	275	4929
Singapore	419	5126	116	1902	81	1342
UAR	213	2311	1453	18110	453	6876
USA	670	11154	332	6550	210	5256
UK	556	9528	418	8083	345	8256
Total (including others)	15138	186720	8891	120457	5570	93309

Source: DG of CI and S, Calcutta.

which can cater the needs of the domestic as well as the external market.

Eventhough, the use of most spices in medicine has declined substantially in recent years, fenugreek is an exception to this. The studies conducted in England indicate that fenugreek seeds substantially contain the steroidal substance diosgenin which is used as a starting material in the synthesis of sex harmones and 'oral contraceptives'. Diosgenin is at present obtained mainly from the tubers of certain species of Dioscorea grown in Mexico and Central America. This is a large duration and costly crop whereas fenugreek is a short duration and non-expensive crop. So, what is needed is, plant breeders should make efforts to develop fenugreek varieties with a high content of diosgenin. This little known spice could make a two-fold economic contribution to a solution of the world's population problems by assisting in birth control and at the same time providing additional food. Apart from this, the industrial use of it in the manufacturing of hair dye, perfume etc., is increasing in recent years. So what is now needed is to make efforts and the implement the policy to improve this sector.

14

NUTMEG AND MACE

Nutmeg is the dried seed of the peach-like ripe fruit of Myristica fragrans—the evergreen tree, native of Moluceas and cultivated in Indonesia, West Indies, Sri Lanka etc. Indonesia is one of the largest suppliers of nutmeg to the world market. In India, it is grown on a small scale in Tamil Nadu, Kerala, Assam and Karnataka and nowadays it is receiving increasing attention among farmers in these states.

Nutmeg and mace are the two distinctly different spices produced from a single fruit of an evergreen, aromatic nutmeg tree usually 9-12 metres high, however, it may grow beyond this also. Mace is the dried reticulated aril of nutmeg. When the peach or apricot like nutmeg fruit bursts open, the mace is seen as an attractive bright scarlet cage closely enveloping or clothing the hard, thin, black shining shell of the seed called nutmeg. The mace is skillfully removed, gently pressed, flat, dried and is called the typical 'blade of mace'. On drying, the original scarlet colour of the mace turns rather pale yellowish brown or reddish brown and becomes brittle. Mace is much more expensive than nutmeg.

Classification of Nutmeg for Trade

1. Whole, sound nutmegs, which consist large, medium and small. These nuts are of interest to the spice trade but the price is too high. The USA and UK import mainly these.

2. Sound shrivels, these are employed for grinding and are too expensive for oil distillation.

3. Rejections, these are low priced and can be used for oil distillation. Germany along with the highest grade consumes this type also.

4. Broken and wormy, this is the cheapest grade, the major consumer of this are European countries which use this in oil distillation.

Types of Mace

(1) 'Banda Mace' considered to be the finest. It has a bright orange colour and a fine aroma.

(2) 'Java Estate' mace is golden yellow, interspersed with brilliant crimson streaks like Banda mace, it is free from insect infestation.

(3) 'Siauw mace' is a lighter colour than Banda mace and contains less volatile oil.

(4) 'Papua mace' often regarded as the fourth grade mace, which contains comparatively little volatile oil.

Marketing

There is no organized market for this product at the producer's level. Most of the producer's use to take this to a distant place for it's sale. Being a commercial crop with the whole production marketed, returns from its cultivation, is mainly dependent on the prevailing market prices. Lured by the steady prices and thus reasonable returns in recent years, a number of farmers in the state of Kerala and Karnataka are shifting towards this crop.

Although nutmeg is giving alternative profits in the recent years, instability in prices, and the consequent risks is the major constraint limiting nutmeg in the long run.

Uses

Nutmeg and mace are generally classified as baking spices, since both are particularly good in sweet foods. Both of these are used as condiments and in medicine. In eastern countries, they are used more as a drug than as a condiment. Nutmeg and mace are stimulant, carminative, astringent and aphrodisiac and are used in tonics and electuaries. In the preparation of medicine for dysentery, stomach ache, flatulence, nausea, vomiting, malaria and early stages of leprosy. However, excessive doses have a narcotic effect.

The oil of this is employed for flavouring food products and liquor. It is used in scenting soaps, tobacco and dental creams and also in perfumery. The oil is official in I.P. It is mildly counter-irritant and used in liniment and hair lotions. It has bean recommended for the treatment of inflammations of bladder and urinary tract.

The butter of this is used as a mild external stimulant in ointments, hair lotions, plasters and it's application is useful in cases of rheumatism, paralysis and sprains. The pericarp or rind of the ripe fruit is locally used in pickles and in the preparation of jellies.

Import of Nutmeg and Mace

Both nutmeg and mace are in great demand in India, but the internal production is too small, which is insufficient to meet the domestic requirements. So India import both of these mainly from Indonesia, Singapore and Sri Lanka. The total volume of nutmeg import in 1997-98 was 293 tonnes valued at Rs. 14883 thousand, this was 508 tonnes in 1996-97 and in 1995-96 it stood at 1269 tonnes as can be seen from the Table-I. As far as mace is concerned it was 351 tonnes in 1997-98. Apart from these Indian import nutmeg oil from the external markets. In 1994-95 the volume of it imported was 20605 kgs. which came down to 4633 kgs. in 1995-96 and again increased to 5511 kgs. in 1996-97 and a fall was observed in 1997-98 to 3960 kgs.

The above aspects clearly show that their appears to be an increasing demand for nutmeg and mace as well as for its oil in India. Instead of spending the much needed forex for this, there is an urgent need to grow it internally in an increasing level. In this regard, area under nutmeg has to be extended gradually in a planned manner. This should be made on a large scale because the prevailing system of cultivation of this in the states of Karnataka and Kerala shows that most of the growers are cultivating it in a small scale, this actually created several problems like improper methods of cultivation, non-availability of better marketing facilities etc.

TABLE-I: Country-wise Import of Nutmeg and Mace into India

(Qty. in Tonnes, Value '000 Rs.)

Country	1995-96		1996-97		1997-98	
	Qty.	Value	Qty.	Value	Qty.	Value
NUTMEG						
Indonesia	886	30248	253	9548	62	2442
Singapore	149	4680	45	1991	25	1729
Sri Lanka	182	6687	210	7767	191	9704
Total (including others)	1269	43447	508	19305	293	14883
MACE						
Indonesia	463	19278	287	15480	224	17351
Singapore	199	5817	149	8099	12	928
Sri Lanka	7	1437	66	4025	88	4656
Total (including others)	609	26532	501	26705	351	24061

Source: DG of CI and S, Calcutta.

So the Government along with the research departments should prepare a long-term strategy for increasing the area under this so as to obtain self sufficiency. The industrial use of it has been increasing in recent years, in the whole world market, hence efforts are needed to increase the production and to start our business in external market. As India is a land of spices, the steps taken to promote this much valued spice is insufficient. Hence, an ideal approach is the need of the hour to this sector. Again, the strategy towards this end should include inter alia measures to stabilise prices at levels necessary for the overall development of the industry.

15

SAFFRON

The Crocus Sativus, a native of Southern Europe and Asian Minor, is one of the oldest and the world's most expensive spices. It is cultivated in Mediterranean countries, particularly in Spain, Australia, France, Greece, England, Turkey, Persia, India, and China. Spain is the world's largest producer of Saffron, accounting for about 90 per cent of the world production.

In India it is cultivated in Jammu and Kashmir and U.P. The valley of Kashmir is famous for its saffron fields and it occupies a major portion of the total area under this in India. A small area under saffron can also be observed in the Kishtwon region of Jammu, Bisar, Doonda and Chanbattia in Uttar Pradesh. The total area under saffron in India at present is around 3800 hectares of which nearly 85 per cent is in Kashmir alone. The present production is around 10 tonnes. In 1970-71 the total area under this was 1628 hectares and the production was 3.90 tonnes. Year after year the total area as well as production has expanded in India, the highest production of 20.04 tonnes was observed in 1979-80 through an area of 3405 hectares since then eventhough the total area under saffron is increasing the production is around 9 tonnes.

Saffron is a low growing plant with an underground globular corm. It is cultivated for its large, scented, blue or lavender flowers. The flowers have trifid, orange coloured stigmas which along with the style-tops yield the saffron of commerce.

Qualities of Saffron

The value of saffron depends mainly on the method by which the stigmas are dried. In Kashmir the stigmas picked from the flowers and dried arranged and constitute "Shahi Saffron", the finest quality which is obtained from the red tips. Here, the flowers are dried in the sun three to five days, then lightly beaten with sticks and passed through course sieves. The material which passes through is thrown into water, those parts of the flowers which float are discarded and the parts which sink to the bottom are collected and further dried, constitute the second quality "Mogra Saffron". The discarded parts of the flowers are again subjected to the beating process and the process of throwing the entire founded mass in water is repeated. The product which sinks is collected, and is very much inferior in value is the third quality called as "Lochha Saffron".

Uses

Historical records shows that, Babar's daughter, Gulbadan Begum, writes about Kesar in *Humayun-nama* and goes in detail to describe the royal mean which included among other delicacies, rice flavoured with sheep's head, curd and saffron. Royality in many countries is believed to have bathed with Kesar mixed water.

Saffron is famous for its extra-ordinary, medicinal, flavouring and colouring properties. It is used abroad in exotic dishes particularly in Spanish rice specialities and French fish preparations. It is also used in fine bread in Scandinavia and in the Balkans.

Apart from its cosmetic uses, it is used for colouring medicines and food to which it imparts a characteristic flavour. It has an aromatic odour and is one of the most popular ingredients used to flavour dishes. It can also give a pleasant colour to butter, cheese, puddings and confectionery.

Saffron is credited with various medicinal properties. It is used occasionally in exam-thematous disease to promote eruptions. It is used in fevers, melancholia and enlargement of the liver and spleen. As it has stimulant, and stomachic properties and is considered to be a remedy for catarrhal infections of children. Saffron is an important ingredient of the Ayurvedic and Unani systems of medicine in India. It is popularly known as a stimulant warm and dry in action, helping in urinary, digestive and uterine troubles. If soaked overnight in water and used with honey, it enables the patient suffering from urine trouble to pass urine freely. Its oil gives strength to the heart and brain, but only when administered in large doses.

Exports

India use to import saffron in the early years, however it entered into the external market in a small way. In 1993-94 we exported about 7 tonnes of saffron to Spain and France. The major markets for our saffron are Spain, Italy and France etc. In 1994-95 the volume of exports came down to 4 tonnes and it was 8 tonnes in 1995-96 and in 1997-98 it was 7 tonnes valued at Rs. 25003 thousands. Eventhough, there is a vast scope to improve the exports from India, there appears to be the prevalence of certain problems in this sector, they are:

1. The outflow of saffron from the production centre to the consumption centres has been paralysed on account of the disturbances in Jammu and Kashmir.

2. Non-availability of quality seeds actually minimized the scope of the growers to increase the productivity and production of saffron.

3. Traditional methods of farming is responsible for lower yields. The yields of saffron, in India is much more than those in other countries.

4. Higher level of tax on this trade.

5. Increasing level of labour and other costs. For example, one pound of saffron consists of about 2,25,000 to 5,00,000 dried stigmas, and it requires the picking by hand itself, of 75,000 flavours. That gives an idea of labour cost involved in harvesting saffron.

6. The good quality of saffron's price is high because of the pre and post-harvesting costs and also the lower yield per hectare.

7. The drying or toasting of the stigmas to obtain saffron leads to an increase in the cost, therefore the price quoted for it is high and it has minimized our scope to enter into the world market in a big way.

8. Adulteration is a major problem. Because of its high cost, saffron is frequently adulterated with styles, anthers and parts of corolla of saffron. So as to increase the weight, water, oil or glycerin is added to this.

The above problems of this sector demand our immediate attention in a proper and planned manner. First of all, there is the need to supply qualitative high yielding seeds along with proper education and training to the farmers. Along with this, there is the need to improve the post-harvesting methods on modern lines. The Government should provide certain liberalised incentives to saffron growers so as to increase its production and exports. Again the growers and traders should concentrate on to supply quality saffron to the market. This will be useful to provide a competition to other competitors in the external market. As a whole, what is required is to extend its cultivation in hilly areas and obtaining both an increase in the production and productivity.

16

VANILLA

Vanilla is an orchid grown in tropical areas. The vanilla pods or sticks are the cured fruits or beans of climbing orchid V. fragrans or V planifolia. It is indigenous to the Western Hemisphere and was found in the 16th century in the coastal forests of southern Mexico. The Spanish brought vanilla to Spain from where its use spread to other parts of Europe notably England and France. Mexico was the leading vanilla producing country of the world for more than three centuries and attempts were made to grow vanilla in tropical Far Eastern countries during the beginning of the 19th century. Thereafter, its cultivation developed outside Mexico, reaching the islands of the Indian and Pacific Oceans.

Vanilla is now cultivated in many countries like Madagascar, Mexico, Camaro on Islands, Tahiti, Indonesia, Zanzibar, Brazil, Seychiles, Malaysia, Sri Lanka, Latin America, Uganda, Tongo etc. India is also emerging as a major producer of vanilla. At present Indonesia and Madagascar are the major producers as well as exporters of vanilla to the world market.

The history of introduction of vanilla to India is rather obscure. It is believed that vanilla was introduced to India about 200 years ago for planting in the 'Spices Garden' at Kourtaliam in Tamil Nadu owned by British India Company. However, the "Wealth of India" tells that vanilla was introduced only 100 year ago. The first attempt in a scientific manner to cultivate vanilla in India was made by Kallar Fruit Research Station in the Nilgiris and Ambalavayal Horticultural Research Station Wayanad.

At present vanilla cultivation is carried out in certain regions of Kerala, Karnataka, Tamil Nadu, and Andamans. The area under cultivation has increased steadily since 1990. The area which was only a few hectares in the initial stage has increased to over 700 hectares in 1997 and reported to reach 1,000 hectares at present. However the production of vanilla in the country is insignificant, which is likely to be between 3-5 tonnes. Production is likely to go up considering the area under cultivation, since one hectare of vanilla can yield 250-300 kgs. The present area after reaching the full bearing state that is 5-7 years may produce around 300 tonnes per annum.

Varieties

There are more than 50 species of vanilla, but only three among them are commercially important as source of vanilla flavour. They are vanilla planifolia Aarndews., V. tatitensis J.W. Moore and V. pampona Schiede. V. planifolia is the cultivated species of Mexico, Cameroon, Reunion Islands, India and Java. V. tatitensis is cultivated in French Polynesia and V. pampona comes up in West Indies. Vanilla beans from different sources have different flavours. While V. planifolia has rich tobacco like somewhat woody deep balsamic and sweet flavour. V. tatitensis has almost perfumery and sweet flavour. The flavour of V. pampona is weak with low vanillin content but has the peculiar sweet floral fragrance.

Vanilla bean is the fruit of an orchid and no other parts of the plant possess flavour character. The typical flavour of vanilla is missing in harvested fresh green beans, which is developed during step by step curing process. The quality of cured beans is determined by its appearance, fragrance, moisture content and to a certain extent is suppleness.

Uses

Vanilla, today, constitutes the world's most popular flavouring for numerous sweetened foods. Vanilla essence extracted from the cured beans of vanilla, is largely used for flavouring ice-creams, chocolates, bakery products, puddings, liquors and perfumes. The demand for vanilla flavour is increasing recently because it used in the preparations of lobsters and praws and the essence is also used for flavouring cashew nuts and ginger products, fruits etc.

Export

Till recently, India used to import vanilla, vanillin, Ethyl vanillin and Iso-Engenol which is used partly for the manufacture of vanillin. It is only in the ninties India started to export vanilla and vanilla concentrate to the external markets. The total volume of vanilla export in 1998-99 was 0.54 tonnes valued at Rs. 1.26 million and this was 0.61 tonnes in 1997-98 and 0.68 tonnes in 1996-97 and in 1992-93, it was 0.40 tonnes. As far as vanilla concentrate's export is concerned it was 0.52 tonnes in 1997-98 valued at Rs. 2.12 million and it was 0.55 tonnes in 1997-98 and the maximum of 1.07 tonnes valued at Rs. 4.53 million was observed in 1996-97. Experts say that there is a big market for vanilla essence both in the domestic and external market considering the growing international market and the possibility to use natural vanilla essence in India, since in India almost all the vanilla essence is based on synthetics, prospects for vanilla bean Industry are rather bright. In this regard, some of the basic issues of this sector have to be solved. They are:

1. Expansion of area under this valuable crop in suitable localities.

2. Educating and training the farmers in various aspects of vanilla farming such as hand pollination vs harmone spray, curing and processing of vanilla beans, their packaging and storage.

3. Encouraging the farmers to start "Vanilla Trust" which can educate the farmers on various pre and post-harvest aspects. For example, in Karnataka in the Dakshina Kannada District in Hiriyadka this type of trust has succeeded in educating the farmers and it is providing the required assistance to them by supplying Vine cutting and in the marketing aspects too.

4. Vanilla development, research and marketing management needs better ideas, planning, integration and outlook for achieving better productivity, performance, service and profits in the industry.

The world demand for vanilla by 2010 will be 71,400 tonnes and in this regard there is vast scope for this crop in India. Again the prevailing demand for this Orchid spice and the future prospects clearly shows that there is ample scope for extending the area in India. In this regard the efforts of the Spice Board by initiating programmes for its developments is an encouraging one. The Board implemented a few schemes for area expansion through distributing vine cuttings and tissue culture plantlets. The department of science and technology and Bio-technology supported the Board for demonstration and increasing the area under cultivation. The Boards programmes like Vanilla New Planting programme, Vanilla Certified Nursery Scheme, Scheme and Scheme for Assisting producers for promoting exports of organic spices etc., if all of these are implemented properly then the growers can extend the area under this crop in a proper manner and the production may climb up and the exports too. Hence, an Action Plan is needed to expand area under vanilla at least to 3500 hectares within a short period so as cater the domestic as well as the external demands.

Apart from all these, presently synthetic vanillin is obtained from coal tar or sulphite liquor from paper mills. Vanillin can also be produced from clove oil Eugenol. However, synthetic vanilla flavour seeks to reproduce only vanillin and is devoid of secondary aromatic compounds such as aldehydes, alcohols and yeast which are responsible to enhance the aroma. Vanilla beans have a delicate aroma with a pleasant after-taste. Despite the advantage, world over, synthetic vanillin is extensively used because of its low price. With the change over to nature friendly products as against those produced from chemicals, there has been a shift in the liking for vanillin produced from vegetative sources. With increase in health consciousness in industrialized and developed countries, the demand for vanillin from vegetative sources is on the increase and the product will also command a higher price.

17

PRICE BEHAVIOR OF SPICES IN INDIA

The price of the spices both in the domestic markets as well as in the external markets are mainly determined by the market forces viz., demand and supply. The supply of spices are determined by the natural conditions, production etc., while the demand in the internal market is decided by the consumption habits and disposable income within the consumers. In the external market the consumption pattern is mainly determined by the quality of spices. In recent years, quality of the produce has emerged as the main criterion in the international trade.

In a developing country like India the consumer's concern over the quality of spices is much lower as compared to that of the developed markets. Because of this the Indian spices are not in a position to obtain a major share in the international market. In the international market for spices various quality parameters viz., physical, chemical, microbiological, sensory and others like insect filth, aflatoxin, pesticide residue, contamination from packaging etc., are adopted while buying the spices. So the price of the spices in the external market is basically noted on the basis of its quality, in this regard the implementation of quality system can result in improvement and sustenance of quality, better image of products, satisfaction to the discerning customer and an attractive price for the product.

Major Wholesale Market Centres for the Spices in India

(1) *Black Pepper*

The wholesale market centres for pepper in India are located mainly in the state of Kerala; there is Kozhikod which handles Nadan and Wynadan grade. While ungarbled pepper is marketed through Cochin centre, Alleppey is famous for Palai and Mumbai is the main export market centre where Malabar grade is handled in larger volume.

(2) *Small Cardamom*

Vandanmethu, Santhanpara, Kumily, Cochin and Calicut are the main auction centres for Cardamom small in Kerala, while Karnataka, Mercara, Siddapura, Sakaleshpura, Sirsi, Mudigere, Yadally are the main auction centres and Pathiveeranpalli and Bodinayakanur are the two auction centres for small cardamom in Tamil Nadu.

(3) *Large Cardamom*

Large is mainly marketed through Siliguri centre.

(4) *Ginger*

Kozhikod and Cochin in Kerala state handle unbleached dry ginger while Mumbai handles bleached dry ginger.

(5) *Chillies*

The main market centre for SI grade of chillies is Tuticorin and for Barik it is Hyderabad for Ist sort Guntur, while Byadgi in Karnataka, Kaddi grade is popular. In the Mumbai centre, Byadgi is handled whereas in Delhi Centre, Guntur grade is marketed.

(6) *Turmeric*

Erode finger grade is popular in Erode, while Rajpuri is handled by Mumbai, Nizamabad grade is marketed in Delhi.

(7) *Coriander*

The major market centres for coriander are Chennai, Tiruchirapally, Ramganj and Mumbai; they handle Guntur, Ist sort, Mandi bold and Indore grade respectively.

(8) *Cumin*

Unjha and Mumbai are the two main wholesale marketing centres for cumin in India.

(9) *Fennel*

For fennel Mumbai and Unjha are the wholesale market centres in India.

(10) *Fenugreek*

Fenugreek sales are undertaken in the Unjha and Mumbai centres on a large scale.

(11) *Celery*

Mumbai is the sole marketing centre for celery in India.

(12) *Nutmeg, Mace and Clove*

For all of these spices, Cochin is the lone market centre in India.

Average Domestic Prices of Spices in India

The average domestic prices of spices in India since 1997 upto 1998 are given in Table-I, which show that the price of black pepper is increasing slowly eventhough there appeared a negative trend in 1998, while for cardamom it increased in negative trend in 1998, while for cardamom it increased in right direction. The same type of tendency as that of cardamom can also be observed in the case of nutmeg and mace along with clove. Except that of turmeric and coriander, almost all other spices have been going upwards in these years. The price index of spices for the year 1997-98 also indicates the same thing.

Price Trend

The wholesale price index of spices and condiments during the year 1997-98 (July-June) generally ruled higher excepting slump in Jan.-Feb. 1998. The index number of wholesale prices which stood at 1,013.6 in July 1997, increased to 1,292.2 in June 1998 making thereby a hike of 27.5 per cent. (Table-II)

TABLE-I: Average Domestic Price of Spices

(Rs./kg)

Spice	Centre	Grade	1997	1998	1999
Black Pepper	Cochin	Ungarbled	200.13	196.00	203.43
		Garbled	206.88	203.00	213.43
Cardamom Small	Vandanmethu	Auction	286.93	323.46	542.44
	Kumily	Auction	284.32	307.63	565.27
	Bodinayakanur	Auction	260.50	312.94	549.26
	Sakaleshpura	Auction	278.06	—	457.20
Cardamom Large	Siliguri	Badadana	79.00	75.81	179.38
		Chotadana	67.60	67.25	159.38
Ginger (Dry)	Cochin	Unbleached	51.25	52.33	79.13
		Bleached	50.25	50.33	76.75
Turmeric	Cochin	Allepey finger	40.00	59.75	41.63
	Mumbai	Rajpuri finger	43.50	45.30	44.00
	Sangli	Rajpuri finger	25.84	26.02	—
Chillies	Tuticorin	—	26.56	30.75	41.38
	Virundhunagar	—	24.63	33.00	34.38
	Guntur	—	17.56	19.50	31.93

TABLE-I: Continued

(Rs./kg)

Spice	Centre	Grade	1997	1998	1999
Coriander	Mumbai	Indori	24.75	27.50	16.31
		Kanpur	26.25	30.70	15.00
Celery	Mumbai	—	19.75	19.50	26.00
Cumin	Mumbai	—	50.75	54.69	80.00
	Unjha	—	44.91	45.21	56.72
Fennel	Mumbai	—	29.38	31.25	56.25
	Unjha	—	19.97	28.15	38.79
Fenugreek	Mumbai	—	12.88	14.88	20.50
Garlic	Mumbai	—	20.00	23.00	25.00
Nutmeg (with shell)	Cochin	—	45.00	70.70	109.38
Mace	Cochin	—	344.00	594.00	517.50
Clove	Cochin	—	93.00	90.00	175.00

Source: Various Issues of Spices India (Kannada).

TABLE-II: Index Number of Wholesale Price of Selected Spices

(Period: July 1997-June 1998) (Base 1981-82=100)

1997-98	Spices and Condiments	Black Pepper	Chillies	Turmeric	Ginger
July	1013.6	567.7	336.0	971.0	676.6
August	1174.1	562.8	332.1	1042.0	732.4
September	1202.4	565.4	333.9	998.7	681.1
October	1305.3	555.8	331.9	988.7	639.1
November	1313.7	555.0	345.1	1012.8	634.5
December	1346.5	570.8	353.6	1056.0	636.5
January	1185.7	588.9	364.5	1150.9	581.5
February	1165.4	589.9	377.1	1330.5	582.5
March	1227.3	592.4	392.3	1295.0	644.1
April	1274.9	604.6	424.6	1260.6	677.1
May	1281.8	603.4	405.2	1241.5	634.2
June	1292.2	615.5	423.1	1248.8	627

Black Pepper

The wholesale prices of black pepper during the year 1997-98 (July-June) generally ruled higher excepting first quarter when it showed a lower trend. The index number of wholesale prices which stood at 567.7 in July, 1997 declined to 555.0 in Nov. 1997 and thereafter the index started increasing and continued and touched 615.5 in June, 1998 making thereby a hike of 8 per cent during the crop year.

Chillies

The wholesale price index of chillies during the first quarter (July-Oct.) 1997 ruled higher upto June 1998. The index number of wholesale prices which stood at 336.0 in July 1997, was recorded higher by 25.9 per cent in June 1998 at 423.1.

Turmeric:

The wholesale price index of turmeric showed a mixed trend during 1997-98, however, the index number of wholesale prices in June, 1998 was recorded substantially higher by 28.3 per cent at 1,248.8 as compared to 973.0 in July 1997.

Ginger

As regards the wholesale price index of ginger the prices were kept fluctuating during the year 1997-98. The index number of wholesale prices which was recorded at 676.6 in July 1997, increased to 681.7 in Sept. 1997 and thereafter moving in both ways, declined to 627.0 in June 1998 making thereby a fall of 7.3 per cent during the crop year.

18

SPICES EXPORT : PRESENT STATUS AND FUTURE PROSPECTS

Spices have profound influence on the course of history and civilization. Eversince the dawn of modern civilization, the Indian Spices have lent flavour and taste to human food, the world over. It was the Europeans' lust for spices that led the Portuguese navigator Vasco-da-Gama to find a sea-route to India, in AD 1948 when he landed at Calicut. For centures, until recently, India held a virtual monopoly in spices production and export.

Consumption of spices varies from one country to another and is influenced to a large extent by the size of the population and the rate at which it grows. It is also influenced by the disposable income, which in the case of developing countries is a major factor. In the developed countries, spices are used in the industrial sector, principally in food processing whereas in developing countries, spices are mainly consumed in individual households. However, the social habits, particularly those of cooking and eating, determine an overall per capita consumption of spices in both industrialized and developing countries. Today, spices are used not only within the house but also in cosmetic, perfumery and indigenous medicine. Now, spices are considered as the natural and necessary component of daily nutrition. Hence, they are used widely not only in India but also in almost all of the countries of the world.

The export of spices is important to India as a means of earning foreign exchange. India exports entire gamut of spices. The spices exported from India include Pepper whole, Dehy. Green pepper, Pepper powder, White pepper, Cardamom small and big, Chillies, Chillies powder, Ginger, Ginger powder, Turmeric, Turmeric powder, Curry powder, Coriander seed and powder, Cumin seed and powder, Celery seed and powder, Fennel seed and powder, Fenugreek seed and powder, Garlic, Dehy. Garlic, Garlic powder, Cassia and Cassia powder, Nutmeg and Mace, Aniseed, Tejpat, Tejpat Powder, MISC spices, oils of spices and Oleoresins of spices.

These products are mostly exported to East European countries, Central Asian countries, American zone and to the Middle East countries in large volume and to the African and Australian zones in small volumes. USA, GFR, Saudi Arabia, Russia are importing spices in large quantities from India.

Export of spices contributed about 4-6 per cent of the total export earnings through agricultural product export during 1995-98, (Table-I) which was 6.72-8.62 per cent during 1992-95. India's export earnings from spices during 1998-99 was Rs. 1,75,60,220

TABLE-I: Share of Spice in Indian Agricultural Exports in Terms of Value

(Rs. in crore)

Year	Total value Agricultural and Exports	Total value of Spices Exports
1995-96	21138	794 (3.8)
1996-97	24239	1202 (5)
1997-98	23691	1408 (6)

Figures in bracket gives the percentage of the total.

thousand and for the period 1999-2000 the target is fixed at Rs. 1,74,80,000 thousand. During the nineties, an increase in both the quantity and value of exports of spices was observed as compared to only an increase in quantity and decline in value of the exports during late eighties. The annual average export earnings in the eighties have gone up to Rs. 1,960 millions from Rs. 830 million in the seventies. Export earnings from spices have touched new records since 1984-85 and in 1987-88 it touched a peak since 1960-61 that of Rs. 2,980 millions. (Table-II). In terms of volume, exports have recorded a whoping growth of 225 per cent from 70,279 tonnes in 1987-88 to 2,31,389 tonnes in 1998-99. As a whole, exports have gone up from 1,02,170 tonnes valued at Rs. 275.76 crore in 1989-90 to 2,31,389 tonnes valued at Rs. 1,758 crore in 1998-99, and the unit value during same period has gone up from Rs. 25.59 per kg. to Rs. 75.98.

TABLE-II: Export of Total Spices from India

Year	Quantity (tonnes)	Value (Rs. '000)	Year	Quantity (M.T.)	Value (Rs. '000)
1960-61	45653	163960	1980-81	92509	1167638
1961-62	64733	175203	1981-82	68375	925101
1962-63	48340	133730	1982-83	75117	928541
1963-64	51179	152793	1983-84	85835	1116622
1964-65	52855	165465	1984-85	89155	2090224
1965-66	62463	230544	1985-86	74501	2825208
1966-67	51714	278191	1986-87	82827	2819943
1967-68	52195	271702	1987-88	70279	2980803
1968-69	51880	250441	1988-89	99946	2748066
1969-70	43975	344798	1989-90	102170	2757609
1970-71	47906	388200	1990-91	109636	2421442
1971-72	67866	366047	1991-92	142104	3809676
1972-73	51662	305612	1992-93	130734	4186364
1973-74	62793	556052	1993-94	182338	5714401
1974-75	54306	625925	1994-95	155008	6201053
1975-76	61952	727248	1995-96	203398	8044301
1976-77	60957	759810	1996-97	225295	12307177
1977-78	81256	1448203	1997-98	242071	14668160
1978-79	104785	1577285	1998-99	231389	17560220
1979-80	114958	1554652	1999-2000*	217850	17480000

*Target proposed by Spice Board.
Source: Spice Board and DG CI and S, Calcutta.

The shift in emphasis from commodities in whole form to value-added products has given a new dimension to the spice industry in the country. Exports of spice oils and oleoresins have recorded substantial growth from 330.98 tonnes in 1984-85 valued at Rs. 97,043 thousand to 1,672 tonnes valued at Rs. 86.76 crore in 1994-95 to as high as 2,750 tonnes worth Rs. 300.77 crore in 1999-2000. As a whole, India has a global monopoly in spices oils and oleoresins.

As far as the export of curry powder is concerned, it was 4245 tonnes in 1995-96 worth of Rs. 1,75,549 thousand and it went up year after year and for this UK is the major market as can be observed from Table-III. The volume of curry powder exported in 1997-98 was 4647 tonnes which went upto 5,000 tonnes valued at Rs. 2,80,000 thousand in 1999-2000.

The importance of individual spices differs from one importing country to another. In terms of whole form, Malabar pepper, Cochin ginger, Alleppey green Cardamom, Alleppey turmeric, and Sannam chillies are some of India's favoured and preferred spice varieties in the international markets. In most of the markets, pepper is almost invariably the principal spice imported in terms of volume and value. Next in importance is the Capsicum group consisting of paprika, chillies and Cayenne pepper, Nutmeg, Mace, Cinnamon and Cassia feature in imports of spices into industrialized markets of Western Europe and North America. Other significant items are ginger, turmeric and seed spices.

Zonal exports of spices and spice products in terms of percentage, volume and value out of the total shows that nowadays, the Middle East and European as well as American zones are importing more of the spices. As a whole, the share of the EEC still dominates in the imports of Indian spices which is 35 per cent in terms of value. However, the share of the American zone has been declining in recent years.

As a whole, the recent performance of India, in her exports of spices is not at all encouraging, while on the other hand countries like Vietnam, Brazil, Gautemala, Malaysia, Indonesia etc., are emerging as the leaders in the global market for spices. The major reasons for this poor show by India is because of the problems of this sector. These are:

TABLE-III: Country-wise Export of Curry Powder

(Quantity-Tonnes, Value '000 Rs.)

Country	1995-96		1996-97		1997-98	
	Qty.	Value	Qty.	Value	Qty.	Value
UK	1348	44680	1294	550	951	42094
USA	240	12767	310	19349	361	20624
GFR	109	4136	177	6024	312	4505
France	87	3293	231	7153	317	9021
Australia	314	13970	218	12449	161	8462
Belgium	177	5389	274	8748	254	9539
Total (including others)	4245	175549	4639	205652	4647	213384

Source: DG of CI and S, Calcutta.

(1) The domestic production of pepper, cardamom, ginger, turmeric and chillies in recent years is dwindling because of

   (a) Lower productivity of black pepper during the last 25 years, and its production have increased substantially, however, the productivity increase was only 24 per cent perhaps due to the crippling incidence of Phytophthora foot rot disease and/or lack of better management practices. In the case of ginger and turmeric, area, production and productivity had increased substantially over the last 25 years. In case of small cardamom, though the area under cultivation registered a declining trend, yet, both production and productivity have gone up significantly. Large cardamom showed only a marginal increase in area and production. Tree spices area and production remained either static or declined.

   (b) Barring ginger and turmeric, none of the major spices have much moved out of their traditional states during the past 25 years, though black pepper has been gradually extending to Andaman and Nicobar Islands and Pondicherry. North East India is emerging as another major area for ginger production.

   (c) Low productivity of these spices in the country as compared to other producing centuries.

   (d) Inadequate supply of high-yielding disease resistant quality planting materials.

   (e) 'Kokke Kandu' viral disease in cardamom, seedling diseases in pepper and cardamom, stunted disease of black pepper, Rhizome rot of ginger, Nematodes, thrips and root grubs in cardamom, shoot borer in ginger and turmeric.

(2) High cost of production. The cost of production of all these major as well minor spices are high in India. Because of these, India is not in a position to compete with other countries in the international market for spices. China and Thailand had become major exporters of garlic while

Morocco and Bulgaria were exporting Coriander at highly competitive prices. Cumin comes from Turkey and Syria at reasonable prices. Export of ginger also dropped substantially because of the price factor. The Cochin ginger which had enjoyed a premium position in the past, is almost extinct now, the major suppliers of ginger at present are China and Nigeria.

(3) Threat from Vietnam, Brazil etc., in the international market in the export of black pepper, for cardamom from Gautemala.

(4) Seed spices were not able to compete with those from other producers such as Turkey, Morocco and Egypt in terms of price.

(5) Increasing level of domestic consumptions of spices which is above 90 per cent of the total internal production.

(6) Indian spice trade is being affected by the stringent sanitary and Phyto sanitary conditions.

(7) Increasing level of sales tax by different spice producing states has actually retarded the interest of both producers as well as consumers. This has ultimately resulted in large volume of smuggling which means government is going to loose the income.

(8) Traditional methods of marketing and the involvement of intermediaries.

(9) Non-availability of warehouses cum cold storage to facilitate storage and transport of spices.

(10) In the external market, there is every scope for imposing a higher tariff by the importing countries in the near future.

(11) Fluctuations in the price of cardamom reduced the interest of the farmers to increase the area and production and now they are planning to shift to coffee and pepper.

(12) Input cost like frequent irrigation and care, credit burden on small growers, non-availability of timely credit etc., are the major hurdles of this sector.

(13) Low genetic variability in tree spices and vanilla.

(14) Poor quality of exported commodities. Failure of the sector in the field of quality upgradation and maintenance besides value addition.

(15) Lack of interaction between the producers or suppliers, processors, manufacturers and buyers.

(16) Non-availability of the required information in connection with the area, production, demand and supply pattern, price trend etc.

Future Prospects:

Import of spices into a major part of the world markets has been growing and the trend may even continue. In recent years, Germany, Japan and some West Asian countries and Saudi Arabia have shown impressive growth rates. The US and UK which is the world's largest market for traditional spices, has also grown in recent years. Taking into account the global needs for spices, there is the need to overcome the above said problems without any delay. This can be made possible only through a properly planned development strategy. It should have the following basic objectives:

(1) Increasing the area, production and productivity of spices. In this regard, there is the need to supply disease resistant high-yielding varieties of planting materials. There is the need for an exclusive nucleus seed multiplication programme to supply quality planting materials to farmers. Again a large scale production of quality planting materials by adopting latest technologies is the need of the hour.

(2) Research activities have to be conducted so as to solve the problem of major diseases and should be demonstrated frequently. In this regard, a package of programme has to be provided to the cultivators.

(3) Then is the need to identify spices varieties having high production potential and exports quality characteristics. Along with this, rejuvenation of the existing unproductive spices gardens is also required.

(4) Need to educate the farmers so as to motivate them to undertake scientific farming through field demonstrations,

input subsidy, extension support etc. Again farmers have to be encouraged to take up mono cropping by using the accredited varieties with an eye on export.

(5) Need to develop the water resources so as to improve the productivity.

(6) Basic facilities for quality upgradation, production diversification, storage and marketing has to be developed.

(7) Efforts and efficiency are needed to improve the database on area, production, market arrivals, prices etc.

(8) There is the need to produce organic spices since there is a growing demand for organic spices in the external markets. The basic rules of organic production are that natural inputs are approved and synthetic inputs are prohibited. The demand for organic foods could jump from 1 per cent to 5-10 per cent of food sales in major markets in the next few years. The grocery stores in foreign countries feature these products nowadays. To meet the fastpaced demand, Indian exporters need to adapt farming, processing and marketing techniques and build good connections with importers and distributors in Europe, the US and Japan.

(9) Government should re-invest all those funds which it has collected from the sector in the form of cess in developmental activities for spices through a transport process involving all key constituents.

(10) Budget allotment for this sector at the centre and state level need to be stepped up to stimulate productivity.

(11) Efforts are needed to catalogue, develop and find use of various non-flavouring properties of spices and to provide academic and background information on the industry.

(12) Need to institute a farm level grading and quality assurance system including sampling and testing capabilities.

(13) Enhancement of food processing industrial activities is basically required so as to utilize the oil and oleoresins of spices.

(14) Organizing conferences in the micro level so as to have a link between the people who are actually involved in this

industry, since, it is a window from which one can get the themes of changing trends and expectations by the countries or buyers and factors that affect their decision.

Hence, the future prospects for the growth of Indian spice industry lies in focusing on value-added products, consistent drive on quality and ensuring competitiveness through enhanced productivity and production. The hurdles of the industry that of a reduction in the exportable surplus or ups and downs in this because of increasing domestic demand, vagaries of nature, pests and diseases and arbitrary imposition of standards by the importing countries are to be overrided without any delay. Indian spice industry should show that it is the only country in the world which is capable of supplying almost all spices according to the taste and preference of the internal and external consumers. As the future of Indian spice exports looks to be quite bright, what is required here is effective utilization of cultivable area under spices. For this purpose, the Government has to prepare a long-term strategy with sufficient resource allocations in the planning activities. Above all, efforts are needed to exploit newer properties of spices beyond their use in food as flavouring agents. Out of all producing countries, India has the best advantages at its disposal to develop the uses and enhance export of higher value-added products as well as growth in consumption of spices worldwide. If we work and implement these in a proper manner then the sector will provide a huge amount of earnings for the development of the nation, since, Indian spices have superior aroma, flavour and pungency which are liked by most of the importing countries of the world.

19

ORGANIZATIONS IN SPICES DEVELOPMENT

Spices research and development programmes in the post-independent India have been formulated with the primary objective of increasing production and productivity, though space and time bound programmes have been added as and when needed. Research and development needs of spices in the country at present are catered mainly by the Indian Institute of Spices Research (ICAR), Spices Board, various State Agricultural Universities, All India Coordinated Spices Improvement Project (AICRPS), CFTRI Mysore (CSIR), RRL, Trivendrum (CSIR) and the Directorate of Cocoa, Arecanut and Spices Development (MOA) Calicut. In general, all these research and development agencies are functioning with the primary objective of increasing production and export of Spices. However, to face the new challenges in the export of quality, Indian spices, additional emphasis has been given on value addition, disease management, genetic resources, production of organic spices, biotechnology research during the last two five years plans.

Organizations engaged in Research and Development and Export Promotion of Spices

(1) Indian Institute of Spices Research (IISR)

The IISR is in Calicut in Kerala state under indian Council of Agricultural Research. It is the apex institute responsible for research coordination in spices in India in collaboration with State Agricultural Universities. The main objective of the IISR is to improve the production and management technologies. It is supplying the planting materials of new crop varieties and also the biocontrol agents for the suppression of diseases and pests.

(2) Directorate of Arecanut and Spices Development

This is under the Ministry of Agriculture, Government of India. It is the nodal agency at the national level for formulating crop development programmes in association with various developmental agencies of the states. The directorate monitors for the implementation of centrally sponsored scheme in Integrated Programme for Development of Spices.

(3) The Spices Board

It is an autonomous body under the Spices Board Act 1986 under the Ministry of Commerce and was created by the Government of India as the agency to promote export of spices in the world. The governing body of the Spices Board consists of exporters, growers, research scientists and officials of the Central and State Government and also the Planning Commission. The Board is primarily entrusted with the responsibility of export promotion. It also acts as a window for importers to get in touch with exporters, as a contact point for farmers and agricultural development organizations to know the latest trends in market demands and as a coordinating agency between various departments and interests.

Research Organizations Working on Various Spices

(1) Black Pepper

(1) Indian Institute of Spices Research (ICAR) Marikunnu, Calicut-673 012, Kerala.

(2) Pepper Research Station (KAU) Panniyur-670 141, Kerala.

(3) Pepper Research Station (APAV) Chintappaly-531 111, A.P.

(4) Pepper Research Station (U of AS), Sirsi-581 401, Karnataka.

(5) Horticultural Research Station (TNAU) Thadiyankudisi-624 212 TN.

(6) Department of Horticulture, Konkan Krishi Vidyapeeth, Dapoli, Ratnagiri-415 712, Maharashtra.

(7) Central Agricultural Research Institute (ICAR), Port Blair-744 101 A and N Islands.

(2) **Cardamom (Small)**

(1) Indian Cardamom Research Institute, Myladumpara, Kailasanadu-685 553, Kerala.

(2) Regional Research Station, Indian Cardamom Research Institute, Thaidyankudisi-624 212, TN.

(3) Regional Research Station, Indian Cardamom Research Institute, Sakalespura-573 134, Karnataka.

(4) Cardamom Research Station (KAU) Pampadumpara-685 553, Kerala.

(5) Regional Research Station (U of AS), Mudigere-577 132, Karnataka.

(6) UPASI Research Station, Vandiperiyar-685 533, Kerala.

(7) Cardamom Research Station, IISR, Appangala, Mercara-571 201, Karnataka.

(8) Horticultural Research Station (TNAU) Yercand-636 601, TN.

3. **Cardamom (Large)**

(1) Regional Research Station, Indian Cardamom Research Institute, Gangtok-737 102, Sikkim.

(2) ICAR Research Complex for NEH Region, Gangtok-732 102, Sikkim.

4. **Turmeric**

(1) Indian Institute of Spices Research, Marikunnu, Calicut-673 012, Kerala.

(2) HIgh Altitude Research Station (Orissa University of Agricultural and Technology) Pottangi-764 039, Orissa.

(3) Regional Agricultural Research Station (APAU) Jagtial-505 327, A.P.

(4) Department of Spices and Plantation Crops, (TNAU) Coimbatore-641 003, Tamil Nadu.

(5) Department of Horticulture, Tirhut College of Agriculture, (Rajendra Agricultural University), Dholi-843 121, Bihar.

(6) Maharashtra Agriculture University, Kasba Digraj-416 315, Maharashtra.

(7) Spices Research Station, (GAU), Jagudan-382 710, Gujarat.

(8) College of Horticulture, (KAU), Vellanikara, Thrissur-680 651, Kerala.

5. Ginger

(1) Indian Institute of Spices Research Marikunnu, Calicut-673 012, Kerala.

(2) High Altitude Research Station (OU of A and T), Pottangi-764 039, Orissa.

(3) Department of Vegetable Crops (Dr. Y.S. Parmar University of Horticulture and Forestry) Solan-173 230, Himachal Pradesh.

(4) College of Horticulture (KAU), Vellanikkara-680 651, Kerala.

(5) ICAR Research Complex for NEH Region, Barapani, Shillong-793 103, Meghalaya.

6. Tree Spices

(1) IISR, Marikunnu, Calicut-673 102, Kerala.

(2) Horticultural Research Station (TNAU) Yercand-636 012, TN.

(3) Horticultural Research Station (TNAU), Pechipari-629 161, TN.

(4) Department of Horticulture, Konkan Krishi Vidyapeeth, Dapoli, Ratnagiri-415 712, Maharashtra.

7. Vanilla

(1) IISR, Marikunnu, Calicut-673 102, Kerala.

(2) Indian Cardamom Research Institute, Myladumpara, Kailasanadu-685 553, Kerala.

Private Organizations in Post-harvest Technology and Trade

(1) Synthite Industrial Chemicals Kolenchery.

(2) D.V. Deo and Co. Cochin.

(3) Concert Spices, Thrissur.

(4) Indian Spices Associates, 85/1, Nehru Nagar, Puttur, Karnataka.

(5) Indian Pepper and Spices Trade Association, Cochin.

(6) International Pepper Futures Exchange Cochin.

Major Activities undertaken through the Organizations

The first 'Pepper Research Scheme' was initiated in 1949 by the then Government of Madras at Panniyur. It was followed by Cardamom Research Schemes at Mudigere in Karnataka and Pampadumpara in Kerala in 1951. During the period 1951-52, the Planning Commission brought to the attention of Ministry of Food and Agriculture, Government of India, the need for research and development in spices and cashew, so a Spices Enquiry Committee (SEC) was constituted to suggest specific measures to develop production and marketing of these cash crops.

By noting down the importance of Spices and Cashewnut as the important foreign exchange earning commodities, recommendations were made to establish an organization at the First Spice and Cashewnut Seminar held in May 1958 at Mercara, Coorg; at a conference on Marketing of Spices, Cashewnut and Arecanut held at Trivendrum in Nov. 1958; and at the meeting of the Governing Body of the ICAR has ultimately resulted in the formation of an Indian Central Spices and Cashewnut Committee under the aegis of the ICAR so as to deal with all aspects of research, development and marketing of these crops, and coordinate these activities in an effective manner.

In 1961, an independent body called the Indian Central Spices and Cashewnut Committee was registered and in 1963 it decided to review the status of spices in the country. The CPCRI, Kasaragod, Kerala came into existence in 1970 and it conducts research on coconut, arecanut, cashew, cocoa, oil palm, pepper, cardamom, ginger, turmeric and tree spices. The Cardamom Research Centre (CRC) at Appongala, Karnataka, originally established by the then Mysore Government, was handed over to CPCRI in 1974.

Till the end of Fourth Five Year Plan, spices research was confined to evolve certain cultural practices in black pepper, cardamom, ginger and turmeric at a few centres under the Department of Agriculture of the present states of Kerala, Karnataka, Tamil Nadu and Andhra Pradesh. Thirty four time bound problem oriented and crop oriented schemes were operational before the Fifth Plan. During the Fifth Five Year Plan, the ICAR established the Regional Station of CPCRI at Calicut in 1975 to undertake concerted research on black pepper, cardamom, ginger, turmeric, nutmeg, clove, cinnamon and all spices. The Regional Station was upgraded to the National Research Centre for Spices in 1986 by merging the CRC, Appangala. The AICRP on Spices was also shifted to NRCS, Calicut. At present, there are 20 research centres spread over 15 states under the AICRP on spices.

The cardamom Research Institute, Myladumpara under spices Board with its regional stations at Sakaleshpura, Thadiyankudsi and Gangtok conducts research on cardamom, post-harvest technology aspects and export oriented projects. UPASI Research station, Vandiperiyar, Kerala also works on Cardamom.

In the Nineties, the Parliamentary Committee (Rajya Sabha) of the Commerce Ministry, reviewed the research and development activities in spices and recommended for the upgradation of NRCS into a full-fledged Institute and transfer of cardamom research to ICAR. Hence, the NRCS was upgraded into a full-fledged 'Indian Institute of Spices Research' Calicut with effect from July 1, 1995.

The other ICAR Institutes working on spices are CARI, Port Blair, A and N Islands, ICAR Research Complex Region, Shillong, The CFTRI Mysore and the RRL Trivendrum, both under the CSIR also have research programmes on post-harvest technology and value addition in spices. The Central Institute of Agricultural Engineering (CIAE), Bhopal again under ICAR, also have programmes on post- harvest technology of spices.

The amount spent on spices development under the Central Sector Schemes was mere Rs. 15 lakhs during the Fourth Five Year Plan and it rose to Rs. 17.5 lakhs during the Fifth Plan, Rs. 30 lakhs during the Sixth Plan, Rs. 240 lakhs in the Seventh Plan and to Rs. 125 crores during the Eighth Five Year Plan.

All these research institutions mainly concentrate on collection, conservation and cataloguing of germplasm of spices and breeding high yielding and high quality spices through both conventional method and/by biotechnological approaches under spices improvement programme. They standardize the plant propagation methods for the production of true-to-type quality planting materials. Micropropagation protocols have been standardized in many spices. High Production Technology for Cardamom and black pepper is being developed through the efforts of these institutes. So as to supply value-added products, Germ plasm accessions of black pepper, ginger, turmeric and cardamom are categorized based on level of essential oil, oleoresin and pungent principles. Again technologies for the management of pests have been developed by these institutes along with the formulation of Integrated Disease Management.

Developmental Programmes for the Spices in recent years

A 'Centrally Sponsored Scheme on Integrated Programme for the Development of Spices' mainly on black pepper and to a lesser extent on ginger, turmeric and chillies was drawn up and implemented in the last three years of the Seventh Plan at an overall outlay of Rs. 435 lakhs. The development measures taken up were production and distribution of black pepper rooted cuttings of high yielding varieties, establishment of model pepper gardens, distribution of input kits, laying out pepper demonstration plots, rehabilitation of old pepper gardens, seedling production and laying out demonstration plots of three spices and popularization of scientific methods of on-farm processing of spices.

In the Annual Plan (1990-92), the Central Sponsored Integrated Programme for Development of spices was expanded by increasing the financial outlay to Rs. 244 lakhs and Rs. 574 lakhs respectively, providing cent per cent central assistance. In addition to the programmes of Seventh Plan, planting materials, multiplication programme of ginger, turmeric, chillies, seed spices and paprika, special programme for spices development in the North Eastern Region etc., where also taken up.

The Integrated Programme for Development of Spices was further expanded with a financial outlay of Rs. 125 crores in the Eighth Plan to increase the overall production of spices both by area expansion and stepping up productivity, reducing the cost of production and improving the quality of the produce of about 27 commercially important spices grown in the country. Because of these efforts, an overall average annual growth in production to the tune of 8 per cent was achieved during the Eighth Plan period as against 4 per cent in the Seventh Plan and the momentum is being maintained during the Ninth Five Year Plan with a substantial financial allocation from Central sector for spices development.

Thrust areas for the future

The following are the thrust areas in the forthcoming years to improve the production efficiency and withstand global competition in the field of cost competitiveness and export.

(1) Standardization of nursery techniques involving biocontrol agents and mycorrhiza for prevention against possible infestation of Phytophthora foot rot of pepper in the initial stages. Popularization of this among the nursery agencies is to be done.

(2) Popularization of integrated crop management technologies among spices growing community without jeopardizing the existing farming systems.

(3) Identification of suitable areas for organic production of spices to meet the increasing specific exports market demand abroad. Necessary arrangements are to be made for certification at each state as per the international norms.

(4) Identifying the post-harvest problems and developing the required facilities for marketing, quality upgradation, product diversification processing, storage etc.

(5) Conducting surveys so to as obtain realistic estimates on area and production of seed spices as well as the minor spices.

20

CONCLUSION

Spices are aromatic vegetable parts like berries, fruits, barks, flowers buds, rhizomes, bulbs, leaves etc. Majority of the spices are marketed in dried form. The spices are essentially used for flavouring and seasonings which make the food acceptable and enjoyable. The flavour value is due to the volatile essential oil and or non-volatile fractions present in the spice. The combined mixture of volatile oil and resinous portion is referred to as Oleoresin. The volatile oil and the resinous portions contribute to the smell and taste characteristics respectively.

The consumption of spices varies from country to country on the basis of the above said characteristics as well as by the disposable income too, along with the social habits. The global spice trade has undergone major changes in the past few years. The food industry mainly the meat processing sector and food service sectors now account for nearly 60 per cent of the spice trade in the developed countries and the household sector has been relegated to second place. The use of spices in confectionery is small.

Pepper outranks all other spices in household consumption, although nutmeg, cinnolnoh, paprika and vanilla are also widely used. All of these are marketed through retail outlets. In recent years, the importance of retail sales is diminishing because of the increasing sale of ready-to-eat foods, fast foods and restaurant food. Eating habits have become more adventurous, with consumers turning more towards ethnic cuisine and spicy foods.

The "hot trend" in spices is increasing the per capita consumption of spices like pepper, chilli, ginger etc.

The food companies including the snack-food are growing not only in the industrialized markets but also in the developing countries. The cuisine in some of these promising new markets like India and China is complex and uses a variety of spices. So the food companies need to recreate the same taste and flavour profiles to capture such markets. The option open to them lies in the spices which they use and how they are used. Due to the bacteriological problems, oleoresins have assumed greater significance in recent years. Spices in other forms, such as encapsulated spices, are entering the market, which could not only influence the usage pattern but also affect the actual amount of spices used, particularly in the industrial sector. Food technologists of leading companies state that they can apply spice alternatives singly or in blends to meet any taste. The advantages of these alternatives are reliable service, consistent quality and not bacterial contamination. There is every scope to gain ground for these in the industries.

The foregoing discussion on the various aspects of the spices produced in India, their uses both traditional and modern, price behaviour, export prospects and research activities clearly show that there is a bright future for the spice industry in India. For this, our farmers, planners, policy makers, researches as well as traders have to work in a positive manner and their efficiency and efforts in production and export can bring a charm to the country, which will bring India into the driver's seat in the global market for spices.

ANNEXURE-I

Individual Item-wise Export of Spices from India During 1994-95 to 1998-99

(Qty. in M.T., Value in Rs. Lakhs)

Item	1994-95 Qty.	1994-95 Value	1995-96 Qty.	1995-96 Value	1996-97 Qty.	1996-97 Value	1997-98 Qty.	1997-98 Value	1998-99(P) Qty.	1998-99(P) Value
Pepper										
Black Pepper	35127.69	22422.67	23750.50	17945.90	45361.29	39241.77	31502.20	45515.22	31121.69	58969.35
Light Pepper	146.83	35.68	288.50	82.69	82.41	29.46	—	—	—	—
Pepper Pinheads	288.65	51.93	227.40	37.04	383.70	74.29	1932.92	1018.07	1123.00	839.79
Pepper Powder	243.04	164.49	399.74	248.95	260.82	189.17	566.59	785.58	846.82	1302.77
White Pepper	266.25	175.43	143.93	162.83	282.90	372.84	26048	596.03	146.12	
Dehy. Green Pepper	246.49	338.66	271.17	506.07	277.30	587.68	288.55	839.73	210.48	833.16
Fr. Dr. Green Pepper	13.37	95.33	27.76	217.98	23.85	144.86	14.77	147.13	23.96	253.07
Pepper In Brine	708.15	198.57	903.35	300.06	691.71	321.26	808.40	379.02	1015.73	882.92
Green Pepper	177.58	150.58	192.70	111.30	180.41	65.57	208.28	208.71	232.25	255.84
Pepper Long	45.66	30.84	38.53	17.02	348.78	204.93	324.31	146.21	143.84	37.18
Item Total	37263.71	23664.18	26243.58	19629.84	47893.17	41231.83	35906.50	49635.70	34863.89	63811.28
Cardamom (Small)										
Cardamom (S)	256.58'	762.61	507.21	1253.75	226.07	869.61	369.37	1261.60	465.77	2472.15
Cardanom (S) Seed	—	—	19.05	41.89	0.01	0.02	1.00	4.98	0.32	1.43
Cardamom (S) Powder	—	—	0.50	1.33	0.02	0.04	0.04	0.20	9.21	47.62
Item Total	256.58	762.61	526.76	1296.97	226.10	869.67	370.41	1266.78	475.30	2521.20

Table *Continued*

Cardamom (Large)										
Cardamom (L)	1247.00	792.79	1582.26	1116.61	1590.07	1177.57	1615.94	1245.45	1400.69	1169.72
Cardamom (L) Seed	46.31	19.95	94.53	77.46	37.47	31.96	32.25	19.02	22.98	21.16
Item Total	1293.31	812.74	1676.79	1224.07	1627.54	1209.53	1648.19	1264.47	1423.67	1190.88
Chilli										
Chili dry	12813.84	3579.85	43969.44	14973.38	32711.69	13257.70	36024.45	9939.22	48273.33	16402.21
Chilli Seed	—	—	108.97	28.00	299.22	90.35	600.01	137.31	792.05	223.37
Chilli Fresh	116.90	16.86	141.42	17.66	951.14	85.11	303.11	30.10	97.19	20.10
Chilli Powder	7102.06	2096.12	11900.98	4515.09	15904.14	6623.90	14644.99	5687.62	11926.57	4950.29
Capsicum Genus	34.90	12.35	—	—	132.61	63.90	95.81	39.56	37.64	20.31
Other Pimenta	28.66	6.44	44.00	12.04	52.21	24.20	110.95	56.21	126.02	44.83
Item Total	20096.36	5711.62	56164.81	19546.17	50051.01	20145.16	51779.32	15890.02	61252.80	21661.11
Ginger										
Ginger Dry	1878.78	892.48	3361.71	2211.64	5514.43	2884.04	7951.04	4476.17	5631.78	3281.48
Ginger Fresh	9972.79	707.15	15024.34	1609.05	23839.67	2811.06	19306.06	2232.41	2507.88	434.87
Ginger Powder	170.91	73.41	96.73	71.44	382.84	229.30	1010.87	554.14	638.50	348.44
Item Total	12022.48	1673.04	18482.78	3892.13	29736.94	5924.40	28267.97	7262.72	8778.16	4064.79

Table *Continued*

Turmeric										
Turmeric Dry	16727.94	2550.53	16989.61	2618.76	10558.46	2469.67	12707.40	3298.58	13856.33	4483.33
Turmeric Fresh/Bulb	5464.07	757.28	3630.98	536.28	4828.08	1096.57	5677.73	1484.14	11043.41	3200.03
Turmeric Powder	6093.73	1210.16	6429.29	1465.29	7632.28	2278.37	10490.20	3523.78	11622.65	4771.61
Item Total	28285.74	4517.97	27049.88	4620.33	23018.82	5844.61	28875.33	8306.50	36522.39	12454.97
Coriander										
Coriander Dry	9621.39	1535.88	10769.70	2031.44	11808.96	2879.68	22407.01	5984.48	18817.99	4046.95
Coriander Powder	1081.00	257.96	771.47	211.90	765.24	256.90	1326.85	450.21	1866.67	542.04
Item Total	10702.39	1793.84	11541.17	2243.34	12574.20	3136.58	23733.86	6434.69	20684.66	4588.99
Cumin										
Cumin White	4414.10	1975.76	2871.86	1416.26	4024.30	2345.44	11396.14	5755.60	8287.81	4690.52
Cumin Black	1046.90	407.56	901.03	263.17	2223.61	1017.50	4528.37	2135.25	2084.92	1060.00
Cumin Powder	156.97	66.33	98.34	59.89	126.70	74.86	356.64	244.67	349.93	260.39
Item Total	5617.97	2449.65	3871.23	1739.32	6374.61	3437.80	16281.15	8135.52	10722.66	6010.91
Celery										
Celery Seed	2575.48	767.05	2620.38	611.47	3740.76	784.41	3262.96	779.58	3962.93	951.19
Celery Powder	26.00	10.22	57.80	13.65	39.14	17.35	53.90	19.61	27.83	17.89
Item	2601.48	777.27	2678.18	625.12	3779.90	301.76	3316.86	799.19	3990.76	969.08

Table *Continued*

Fennel										
Fennel Seed	2016.61	576.66	2579.12	744.86	4850.07	1788.60	12333.03	3567.10	5221.61	1513.27
Fennel Powder	12.55	4.89	15.05	6.87	—	—	35.30	14.73	57.48	24.84
Item Total	2029.16	581.55	2594.17	751.73	4850.07	1788.60	12368.33	3581.83	5279.09	1538.11
Fenugreek										
Fenugreek Seed	7910.30	1215.43	15108.18	1860.02	8813.92	1181.14	5879.52	955.01	9856.79	1843.47
Fenugreek Powder	45.34	9.54	29.77	7.17	76.73	23.43	126.61	32.14	224.98	71.44
Item Total	7955.64	1224.97	15137.95	1867.19	8890.65	1204.57	6006.13	987.15	10081.77	1914.91
Item Total	2338.20	486.86	2493.63	518.52	3059.09	842.60	4056.16	934.53	2001.52	749.11
Garlic										
Garlic	470.96	53.29	3600.55	346.60	3910.17	463.79	2277.92	166.82	2684.51	229.32
Dehy. Garlic Flakes	95.15	34.73	217.30	96.15	233.94	87.85	658.60	259.78	1116.08	417.44
Dehy. Garlic Powder	67.29	34.84	117.68	48.51	744.93	246.10	1038.09	370.96	267.97	94.26
Item Total	633.40	122.86	3935.53	491.26	4889.04	797.74	3974.61	797.56	4068.56	741.02
Tree Spices										
Clove	35.78	11.07	0.29	0.56	83.10	24.04	2.19	5.29	2.11	6.62
Nutmeg	5.02	5.69	5.85	6.63	3.85	2.82	4.52	6.64	1.76	4.18
Mace	—	—	0.04	0.19	1.40	1.79	0.14	0.56	0.07	0.15
Vanilla	—	—	—	—	0.68	4.00	0.61	3.90	0.54	12.60

Source: DG CI and S, Calcutta.

ANNEXURE-II

Individual Item-wise Export of Spices Oils from India During 1994-95 to 1998-99

(Qty. in M.T., Value in Rs. Lakhs)

Item	1994-95		1995-96		1996-97		1997-98		1998-99(P)	
	Qty.	Value	Qty.	Value	Qty.	Value	Qty.	Value	Qty.	Value
Spice Oils										
Pepper Oil	41.27	353.30	34.73	358.88	36.16	414.93	68.89	1519.79	86.25	2606.12
Cardamom Oil	1.05	61.11	0.31	13.88	0.43	22.07	0.35	24.26	2.13	116.05
Chilli Oil	1.43	19.53	0.25	1.31	0.01	0.08	1.15	23.40	0.22	3.86
Capsicum Oil	0.42	16.05	0.35	3.54	0.67	10.69	1.54	19.22	—	—
Paprika Oil	—	—	3.60	16.44	—	—	5.48	33.88	—	—
Ginger Oil	7.52	108.01	5.26	142.00	4.12	109.56	10.38	346.45	14.73	424.46
Coriander Seed Oil	—	—	0.03	0.24	0.50	8.87	0.35	3.40	3.40	—
Cumin Seed Oil	0.55	13.40	0.71	13.55	2.17	30.70	11.09	19.45	0.50	16.19
Fennel Seed Oil	0.01	0.17	0.00	0.12	0.06	1.76	0.01	0.18	0.06	3.14
Fenugreek Seed Oil	0.25	0.88	—	—	0.00	0.04	—	—	—	—
Garlic Oil	0.13	3.94	0.17	3.48	0.15	3.69	1.07	18.31	0.24	17.53
Clove Oil	6.25	25.81	4.09	19.00	1.58	9.01	2.79	22.37	5.17	29.38
Nutmeg Oil	3.63	13.12	2.90	18.35	2.87	20.31	5.54	44.29	19.11	261.97
Mace Oil	0.23	1.80	0.58	6.25	0.05	0.71	0.32	2.80	2.79	40.16

Source: DG CI and S, Calcutta.

ANNEXURE-III

Individual Item-wise Export of Spices Oleoresins from India During 1994-95 to 1998-99

(Qty. in M.T., Value in Rs. Lakhs)

Item	1994-95		1995-96		1996-97		1997-98		1998-99(P)	
	Qty.	Value	Qty.	Value	Qty.	Value	Qty.	Value	Qty.	Value
Spice Oleoresins										
Pepper Oleoresin	503.74	2507.01	559.51	3501.20	518.57	3240.99	561.85	5566.02	617.60	7342.32
Cardamom Oleoresin	0.21	6.28	1.80	26.82	8.64	134.24	0.33	6.65	0.95	75.85
Chilli Oleoresin	45.94	498.26	14.28	126.77	11.71	165.94	105.00	1063.27	27.52	229.45
Capsicum Oleoresin	227.11	1324.73	195.26	1337.79	268.52	2163.53	259.12	2008.07	290.34	1986.55
Paprika Oleoresin	82.43	774.34	157.39	1734.07	261.51	3187.28	487.98	5818.04	731.52	7865.21
Ginger Oleoresin	57.91	532.94	56.73	547.44	45.31	450.61	67.98	732.84	97.84	1253.70
Turmeric Oleoresin	159.00	1000.63	161.68	855.42	154.69	867.33	199.95	1706.09	240.63	3199.60
Coriander Oleoresin	0.70	3.70	7.14	27.69	13.76	87.52	1.42	12.97	2.31	19.62
Cumin Oleoresin	0.35	5.29	8.95	87.97	5.68	44.65	8.23	79.02	2.00	23.86
Celery Oleoresin	126.85	304.52	133.41	366.25	151.13	363.03	149.88	480.99	144.06	285.44
Fennel Oleoresin	0.01	0.05	4.07	17.24	5.48	12.38	3.67	19.99	6.01	41.67
Fenugreek Oleoresin	2.70	3.44	21.42	92.66	2.23	24.52	0.61	3.95	9.26	52.19
Garlic Oleoresin	0.72	5.04	4.32	34.10	1.69	11.83	2.40	32.77	3.85	49.16
Clove Oleoresin	0.31	2.73	1.64	9.79	1.15	8.08	1.96	36.62	2.21	23.37
Nutmeg Oleoresin	23.52	91.98	22.16	95.93	35.20	157.69	44.21	261.97	47.13	279.48
Mace Oleoresin	3.83	31.70	3.70	41.37	6.77	61.24	24.49	200.21	18.64	307.83
Vanilla Concentrate	0.15	6.62	0.73	28.06	1.07	45.25	0.55	12.97	0.52	21.26

Source: DG CI and S, Calcutta.

ANNEXURE-IV

Country-wise Export of Total Spices from India During 1994-95 to 1998-99

(Qty in M.T., Value in Rs. Lakhs)

Country	1994-95 Qty.	1994-95 Value	1995-96 Qty.	1995-96 Value	1996-97 Qty.	1996-97 Value	1997-98 Qty.	1997-98 Value	1998-99(P) Qty.	1998-99(P) Value
Afghanistan	69.63	52.03	480.29	273.08	158.92	124.44	709.39	365.92	95.51	137.83
Australia	1161.98	647.68	934.16	650.73	1296.01	950.40	1276.71	1265.48	1808.26	2356.30
Algeria	202.00	68.43	195.20	81.73	93.81	26.02	33.00	57.22		
Argentina	226.01	391.34	158.34	303.01	58.50	188.24	428.82	751.69	199.72	291.03.
Austria	3.12	6.23								
Antigua and Barbuda	0.30	0.25	22.23	3.46	24.37	42.11	92.57	116.20	40.28	100.16
Bulgaria	70.25	47.22	76.25	60.23	52.30	32.79	8.70	6.67	4.68	6.05
Baharain	886.57	277.81	1051.11	408.77	790.50	288.31	646.79	314.92	649.58	401.51
Belgium	370.33	140.64	500.65	237.79	829.18	383.99	841.68	841.63	1249.58	1260.07
Bangladesh	8510.82	676.42	21817.27	4349.93	6159.62	778.78	8315.76	975.28	9231.32	1948.76
Burma (Myanmar)	228.15	50.47	3526.88	539.43	1386.17	283.02	60.50	14.40		
Brazil	331.60	672.81	215.44	443.42	289.60	1520.16	1037.70	1217.08	1052.53	1541.52
Barbados	5.00	2.86	38.00	11.58	178.27	80.81				
Bahamas	13.00	10.09								
Bolivia	0.13	0.95								
Botswana	0.20	0.10								

Table *Continued*

Country										
Brunei	9.69	1.94								
Burundi	0.13	0.08								
Bhutan	16.00	5.48	5.24	5.28	5.11	3.75				
Bermuda	5.00	4.73								
Belarus	561.35	471.95								
Brvirginis	12.00	6.45	12.00	4.25						
Canada	3058.52	1727.55	2940.84	2033.23	4187.42	3120.89	4428.68	4407.64	5099.78	6158.30
China	18.36	36.70	306.26	70.02	381.45	140.32	441.52	748.38	299.77	401.05
Cyprus	3.66	1.77	6.60	6.17	20.50	9.97	30.91	31.41	4.43	7.80
Canary islands	5.00	7.67								
Columbia	50.00	18.18	35.00	10.81	98.55	41.22	38.00	57.95		
Czechoslovakia	0.50	0.83	0.47	4.90						
Chile	18.80	28.34	22.11	16.05	69.47	51.00	168.78	133.87	125.31	115.05
Congo	0.22	0.16	0.14	0.09						
Costa Rica	63.00	142.42								
Cuba	0.01	0.09								
Cameroon	1.04	1.48								
Comoros	2.20	1.58								
Channel Islands	3.19	20.81	1.44	7.98	10.00	1.23				
Croatia	76.94	44.77	16.01	3.07	25.00	10.28	27.46	52.79		
Czech-Republic	42.90	19.67	104.98	88.62	194.58	150.24	19.12	23.34	33.53	110.36
Denmark	344.98	232.13	448.98	415.50	383.49	399.41	347.78	709.08	413.44	1454.32

Table *Continued*

Djibouti	54.46	7.95	88.00	18.55	32.15	9.40	70.50	17.20	50.00	21.58
Dominica	2.00	0.82	32.32	13.22						
Egypt (A.R.E.)	2084.58	769.66	1893.37	837.72	1689.47	525.22	2600.58	1258.49	3158.44	2152.30
Ethiopia	25.30	7.98	45.19	13.00	50.35	15.40	129.72	30.32	189.47	59.43
Equatorial Guinea	0.01	0.05								
Estonia	0.20	0.24	16.29	10.51	5.00	0.86				
Finland	6.04	28.36	4.61	10.40	6.34	35.85	162.96	275.35	228.01	464.43
France	3604.49	1812.96	3721.43	2119.89	3033.40	2583.52	3573.92	3881.22	3343.81	4427.51
Fiji	0.05	0.64	71.77	27.66						
Faerose Island	3.00	0.55								
Greece	143.64	84.17	393.90	218.28	366.95	200.15	519.46	288.63	278.84	169.03
Germany	4041.69	2472.90	3889.14	2811.97	5246.22	3894.45	7078.35	6372.09	5375.52	7162.23
Gambia	15.00	21.63	10.00	3.55	0.05	0.46				
Ghana	0.01	0.03	17.64	6.77			3.71	2.79		
Guatemala	20.50	3.61								
Guinea	58.99	22.26								
Guyana	65.30	14.20	65.00	20.20	129.80	50.14				
Gibraltar	0.05	0.06			7.63					
Guadeloupe	43.95	16.00	47.05	14.60		5.14				
Georgia	73.31	117.37	38.34	66.54						
Hungary	10.03	1.47	0.22	1.99	116.67	90.71	4.19	16.37	284.70	593.74
Hongkong	363.65	171.48	759.55	339.56	2602.92	1144.46	3933.89	1950.36	1276.07	895.58

Table *Continued*

Honduras	24.83	9.15	75.00	40.18						
Haiti	0.58	8.66	18.93	5.97						
Iraq	86.50	29.10								
Italy	3168.43	1711.74	3105.09	2301.41	2942.28	2328.27	3075.92	3780.94	2223.66	3370.82
Indonesia	363.29	151.04	273.20	70.66	916.79	303.33	3221.97	872.20	1806.80	481.87
Ireland	3.11	33.90	20.95	75.47	16.20	18.54	140.08	240.26	32.20	105.71
Iran	1509.78	253.72	3511.08	670.40	2470.34	606.91	4031.36	1040.79	1126.95	329.63
Israel	1139.01	287.05	1175.88	427.43	1109.38	468.62	1781.72	946.52	1333.12	760.34
Ivory Coast	1.10	0.48	1.40	0.74	0.54	0.24				
Japan	8338.01	3276.30	8173.31	2687.42	8282.44	7409.56	6618.67	7514.49	6512.19	8845.18
Jordan	501.19	67.13	978.28	118.49	646.54	160.41	417.75	93.86	712.96	200.32
Jamaica	13.00	13.54	1.00	1.21						
Kuwait	1014.94	375.53	1350.83	659.71	1273.36	521.15	1433.84	738.84	1389.47	958.77
Korea (South)	598.57	504.14	669.86	265.44	849.35	656.45	597.21	794.26	407.03	932.96
Korea (North)	121.09	48.98	34.15	345.61	64.14	253.00	12.33	147.27	78.11	259.29
Kenya	426.29	95.51	372.67	68.99	406.50	92.66	281.67	137.38	448.30	221.97
Kazakistan	19.50	14.96	1.21	2.44						
Kyrghyztan	2.65	7.77	12.50	3.67						
Libya	0.01	0.95			1087.50	232.08				
Lebanon	87.84	17.08	133.40	25.94	72.20	21.41	133.75	46.93	214.00	95.08
Liberia	38.00	6.03	71.60	26.93						
Luxembourg	0.09	0.05								

Table *Continued*

Latvia	44.52	72.48	179.80	163.57	946.09	445.76	1206.70	559.09	1793.49	695.89
Lithuania	12.25	5.52	1.00	0.78	85.55	34.06	65.11	35.64	207.72	75.05
Morocco	1194.40	290.86	2588.28	581.04						
Maldives	227.73	55.79	79.14	30.25						
Malawi	22.10	7.82	3.80	2.28	8.05	2.11	6.98	14.00	12.34	5.81
Malaysia	4345.93	885.57	6623.02	1809.43	11807.86	4203.23	9226.80	2851.92	8829.86	2811.66
Mauritius	579.81	143.37	701.38	187.31	884.65	305.21	931.68	356.51	839.26	315.15
Mexico	204.05	276.44	310.53	255.27	1255.73	688.14	3657.35	1735.81	1084.39	2443.83
Mongolia	9.50	2.39			5.99	3.96	3.00	1.64		
Mozambique	134.09	23.81	2.87	1.12	1.11	1.28				
Martinique	6.00	1.64	647.86	113.83						
Macau	0.20	0.13	161.93	58.88						
Mali	35.56	7.23	16.70	4.89	21.10	6.60				
Malagasy Rep	3.16	2.16	14.98	12.69	5.25	1.73	14.00	7.52		
Norway	12.69	2.36	10.44	2.54	8.73	2.34	25.91	8.22	64.41	99.72
Netherlands	3005.87	1296.01	3537.90	1606.22	6186.24	3908.69	4983.58	3151.92	4334.56	4218.48
Nepal	1630.51	697.62	1900.44	1002.52	1926.15	1177.89	2697.33	1361.27	4838.27	2692.69
New Zealand	240.31	119.62	227.87	120.09	361.23	228.67	596.00	424.08	369.32	439.57
Nicaragua	1.04	7.16								
Nigeria	68.55	30.22	17.58	17.30	15.97	15.86	0.93	10.18	127.99	111.21
Niger	11.00	12.34								
Netherland Antils	0.98	10.93								

Table Continued

Oman	546.87	172.31	566.87	218.86	493.08	187.17	583.30	303.51	454.85	277.04
Panama	29.01	9.64	117.04	39.39	339.19	135.06	183.75	93.28	67.30	34.83
Portugal	102.53	19.03	1335.95	1025.25	966.34	836.56	1147.34	1166.04	834.69	1324.17
Poland	2049.60	1526.69	20406.95	3260.50	26920.22	4221.81	22078.16	3540.69	27377.03	5673.91
Pakistan	6827.36	1060.72	20.70	25.95	44.80	65.73				
Paraguay	19.71	14.21	96.76	52.25	20.69	61.62				
Peru	0.04	0.55	106.79	122.09	461.93	204.64	394.81	369.46	780.06	590.73
Phillipines	43.76	66.12					261.77	154.09	318.50	153.07
Papua New Guinea	0.14	0.03	Neg.	Neg.						
Qatar	129.03	44.36	377.32	121.94	157.89	51.58				
Romania	101.90	75.19	171.07	110.04	29.96	30.75				
Reunion	13.45	8.91	5.65	2.42	6.89	5.04	0.30	1.53		
Rwanda	7.75	2.61	8.25	33.14	2.56	4.20	51.80	29.75		
Russia	6516.38	4729.03	3985.77	2990.72	3861.09	3226.26	4533.67	5496.60	2108.95	3506.40
Singapore	5568.69	1518.61	8707.19	3188.82	13813.58	6243.17	11067.69	5344.19	7175.16	6996.69
Sweden	240.75	225.36	252.99	225.71	335.53	301.37	283.51	456.02	311.79	645.27
Spain	787.77	590.86	1116.28	1243.38	2200.69	3176.28	2511.02	3174.77	806.55	1591.49
Switzerland	1001.28	192.04	916.39	193.64	697.71	181.34	589.80	267.05	205.18	240.04
Syria	346.28	67.73	439.55	82.62	508.00	105.36	291.38	71.68	518.59	200.41
Sierra Leone	0.04	0.02	2.22	22.78	6.00	8.98				
Senegal	18.00	3.25	18.00	4.30	36.00	8.29	2.20	1.28		
Seychellis	64.15	37.62	35.69	16.05	26.61	10.04	32.11	17.95	39.77	39.33

Table *Continued*

Somalia	0.55	1.01	33.15	22.99	2.83	1.93				
South Africa	4008.79	907.92	4927.19	1240.47	4145.34	1352.90	7808.46	2993.25	5473.62	2562.72
Sri Lanka	1162.47	2086.79	14989.28	3780.00	10256.30	2685.48	20925.93	4273.86	27507.39	8997.10
Sudan	30.60	16.46	109.98	71.88	50.93	30.86	70.86	111.12	154.62	123.13
Surinam	25.63	8.39								
Saudi Arabia	3394.23	1131.75	5530.13	2711.04	5831.68	2357.09	5424.68	2735.17	6132.64	2782.26
Saint Lucia	13.00	3.08								
Soloman Islands	14.50	13.14								
Slovenia	15.11	11.62	32.41	27.98	135.97	52.06	25.00	50.97	14.08	34.74
Slovak Rep	1.50	1.67								
Thailand	73.10	39.84	323.75	121.88	877.00	389.62	175.58	143.76	107.59	170.54
Tanzania	58.71	11.28	47.03	14.35	25.36	12.03	38.16	26.75	42.49	45.21
Togo	18.00	4.60								
Trinidad	101.67	27.31	61.15	14.11	80.24	24.51	163.12	112.32	313.58	172.37
Tunisia	494.01	112.05	919.00	108.85	591.85	167.98	540.20	174.84	509.48	230.62
Turkey	152.07	25.38	286.81	152.40	343.29	197.90	673.75	355.63	557.94	266.65
Taiwan	645.46	313.69	1437.33	231.25	2255.57	804.73	1127.13	684.11	932.52	1295.59
Tokelao	10.00	16.54								
Turks and Caicos Island	30.82	3.26	0.14	0.10	0.02	0.04	1.02	0.84		
Turkemenistan	28.00	14.10								
USA	29241.56	18303.76	24301.01	18016.670	46872.35	41774.77	42872.28	43768.46	38243.73	53714.81
UK	8329.48	3513.68	8973.11	4729.79	10613.61	6965.83	12234.48	9054.75	11442.69	11355.36

Table *Continued*

Uganda	1.13	0.45	0.44	0.60	19.30	11.58	10.73	4.21	3.54	3.03
Uruguay	0.85	6.53	0.05	0.48	29.16	18.13	28.22	20.35	1.98	13.80
Ukraine	230.48	186.19	221.22	166.97	247.72	186.52	478.02	211.16	198.26	141.79
Venezuela	0.20	2.02	66.50	16.25	46.10	16.45	64.10	51.384	46.24	50.69
Vietnam	0.37	0.39	0.54	0.28	45.72	79.72	31.03	45.15		
West Indies	25.28	19.68	18.00	12.68						
Y.A.R.	1688.61	415.79	1806.61	605.40	3281.64	841.75	1785.13	662.23	2455.75	705.57
Yugoslavia	74.90	48.97	14.98	13.33	11.50	19.46	14.00	26.64		
Zambia	27.26	15.33	5.36	3.29	0.91	0.80	3.83	1.64	0.26	0.16
Zaire	0.36	0.18	3.00	1.78	0.10	0.17				
Zimbabwe	24.47	8.02	85.43	28.27	94.91	41.05	56.32	37.93		
Total	155008.46	62010.52	203398.42	80443.05	225294.61	123071.76	242070.75	146681.64	231389.44	175802.12

(P) Provisional.

Source: DG CI and S, Calcutta/S.Bill/Exporters/Returns.

BIBLIOGRAPHY

Abraham, P.—Spices as intercrops in coconut and Arecanut Gardens. Bulletin of the Indian Central Coconut Committee 10: Kasaragod, 1956.

Anonymous—Spices—Trends in World Markets, Commodities Bulletin Series 34: FAO, Rome 1962.

Anonymous—Know your market for Cardamom, Cardamom Board, Cochin 1974.

Anonymous—Cardamom in Karnataka, University of Agricultural Science, Hebbal, Bangalore, 1976.

Anonymous—Market Survey of Cardamom in Selected Markets in India, Cardamom Board, Cochin, 1980.

Anonymous—Spices: A Study of World Market Vol. I and II, International Trade Centre, Geneva, 1982.

Anonymous—Production and Trade in Minor Spices in India—ICA and S Journal 8 (1) 1983.

Anonymous—Domestic Survey on Cardamom and Pepper, Spices Board, Cochin, 1987.

Anonymous—High Production Technology in Cardamom, National Research Centre for Spices, Calicut, 1989.

Anonymous—Export Promotion of Spices, Spices Board, Cochin,1990.

Anonymous—Spices, 10th Annual Report, Spices Trading Corporation Limited, Bangalore, 1993.

Chandola, R.P., Mathur S.C. and Srivastava V.K. -"Cumin Cultivation in Rajasthan" Indian Farming 20 (4), 1970.

Divakaran—Development of export value of Spice product, J of Spice India 3 (4) : Spices Board, Cochin, 1990.

Directorate of Arecanut and Spices Development, Calicut—Development of Ginger, Turmeric and Chillies in the Fifth Five Year Plan—An Approach Paper 1973.

DMI, Nagpur—Marketing of Chillies in India, Marketing Series No. 83, 1957.

DMI, Nagpur—Marketing of Cardamom in India, Marketing Series No. 144, 1965a.

DMI, Nagpur—Marketing of Turmeric in India, Marketing Series No. 148, 1965b.

DMI, Nagpur—Marketing of Ginger in India, Marketing Series, No. 151, 1966a.

DMI, Nagpur—Marketing of Minor Spices in India, Marketing Series, No. 163, 1968a.

DMI, Nagpur—Marketing of Pepper in India, Marketing Series No. 171, 1971.

Eswara Prasad Y and others—An analysis of arrivals and prices of turmeric in Guntur Market—IJ of AM3 (1), 1989.

George, C.K. Shivadasa, C.R. and others—Strategies for Export Development of Spices Board, Cochin and International Trade Centre, Geneva, 1989.

George, Paul—Spices Industry on the March—Agro India, Feb. 1998.

Guenthur Ernest—The essential oils, Robert E. Pub Company, Hamilton, New York, 1975.

Hemannu, Shashidhara K.—Medicine from Spices J of Spices 5(5) : Spices Board, Cochin, 1992.

ICAR—Condiments and Spices, pp. 1165-1187, Hand Book of Agriculture, 1997 (Reprint).

Jaya, S. Anand—Pepper Marketing by Cooperatives: A case study—Indian cooperative Review 28 (4) 1991.

Khan, M.T.—Spices in Indian Economy, Academic Foundation, Delhi, 1990.

Krishnakumar, V. and Potty S.N.—Vanilla an Orchid Spice—Agro India Sept. 1999.

Lakshmanachar, M.S.,—Pepper Product Development and its Trade—ICA and S Journal 13(1), 1989.

Madhava, K. and Bosco—Post-harvest Technology and Processing of Plantation Crops—Improvement of Plantation Crops (Edi), CPCRI, Kasaragod 1999.

Mahindran, S. N.,—Spices in Indian Life, Chand and Sons Pvt. Ltd., New Delhi, 1982.

Naidu, M.R. and Hanumanthaiah C.V.—Price Spread of Turmeric and Chillies Regulated Marketing in Guntur district, AP, a comparative study— IJ of AM 2(1) 1988.

Naidu, M.R. and Srirammurthy C.—Inequalities in distribution of area and production of turmeric in AP—ICA and S Journal 13(2), 1989.

Natarajan and others—Production of White Pepper, Pepper Oil and Pepper Oleoresin—Indian Spices 1967.

Nambiar, M.C.—Report on the Spices and essential oil crops of India, FAO, Rome, 1978.

Peter, K. V.,—Spices—Silver Jubilee Souvenir—CPCRI Kasaragod 1997.

Peter, K.V.,- Spices Research and Development-Agro India Jan. 1998.

Peter, K.V.,- Spices Research and Development—An Updated Overview—Agro India Aug. 1999.

Pruthi, J.S.—Spices and condiments—National Book Trust, India, New Delhi, 1979.

Pruthi, J.S.—Diversification in Pepper Utilization by development of value-added processed products—ICA and S Journal 13(1) 1989.

Randhawa, M.S.—a History of the Indian Council of Agircultural Research—1929—1979 ICAR New Delhi, 1979.

Shenoy, J.R. and Ravindran TC—Fluctuations in the Prices of Pepper in recent years—ICA and S Journal 13(1), 1989.

Sikka, R.K.—Price Spread in Ginger Trade- Arecanut and Spices Bulletin 7 (3), 1976.

Sikka, R.K. and George C.K.—Price Spread of Ginger—ICA and S Journal 7(1).

Suresh Shah—Indian Spice Girls Record a Hit—The Economic Times, 7th Feb. 2000.

Tiwari, S.C. and others—Ginger cultivation in Himachal Pradesh—An economic analysis—University of Horticulture and Forestry, Solan 1986.

Vigneshwara, V—Pepper and its Problems—Facts for You 9(3), 1987.

Vigneshwara, V.—Pepper in India—Yojana 33(19), 1989.

Vigneshwara, V.—Spices in India—I, Facts For You 11(5), 1989.

Vigneshwara, V.—Spices in India—II, Facts For You 11(6), 1989.

Vigneshwara, V.—Ginger Production : Problems and Prospects—Yojana 34(18), 1990.

Vigneshwara, V.—Pepper Exports: Full of Prospects—ALIVE Nov. (2), 1991.

Vigneshwara, V.—Indian Cardamom Export : Whither the aroma of Yester years—Financial Express July 13, 1991.

Vigneshwara, V.—Spices Export Market: Is the Flavour in India's Favour ?—Financial Express—Sept. 28, 1991.

Vigneshwara, V.—The Golden Spice losing its glitter ? —Financial Express, Jan. 11, 1992.

Vigneshwara. V.—Cloves: Vast Scope—Financial Express. April 25, 1992.

Vigneshwara, V.—Low Yield Plauges Chilli—Financial Express, Aug 1, 1992.

Warrier, P.K.—Spices in Ayurveda—J of Spices India 3(5) Spices Board, Cochin, 1990.

Zacharin John T.,—Post-harvest Processing of Spices—Harvest and Post-harvest Technology of Plantation Crops—CPCRI, Kasaragod 1998.

INDEX